Going Backwards:
Reverse Logistics Trends and Practices

Going Backwards:
Reverse Logistics Trends and Practices

University of Nevada, Reno
Center for Logistics Management

Dr. Dale S. Rogers
Dr. Ronald S. Tibben-Lembke

Contents in Brief

Table of Contents

CHAPTER 1: SIZE AND IMPORTANCE OF REVERSE LOGISTICS　　1

CHAPTER 2: MANAGING RETURNS　　37

CHAPTER 3: DISPOSITION AND THE SECONDARY MARKET　　73

CHAPTER 4: REVERSE LOGISTICS AND THE ENVIRONMENT　　101

List of Tables

List of Figures

Reverse Logistics Executive Council Founding Members

Avery Dennison
Black & Decker Inc.
Braun, Inc.
Canadian Tire Corp.
Cosmair
Egghead Software
Federated Department Stores
GENCO Distribution System
Home Shopping Network
Kmart Corporation
Levi Strauss & Co.
Nike
Nintendo of America
Polo / Ralph Lauren
Sears, Roebuck, & Co.
Sharp Electronics Corp.
Sony Electronics Inc.
Surplus Direct
Target Stores
Thompson Consumer Electronics

Authors Notes

There are many people who have made this piece of work possible. This project has truly been a group effort. Without the assistance of several student research assistants this work could have not been completed. Mihaela Ghiuca, the 1998-99 Reverse Logistics Executive Council Fellowship student; Jacob Brothers, now of Minute Maid; Heather Goulding, now of Skyway Logistics; Jin Hong and Lily Schwartz of Sun Microsystems; Ruohong Zhao of Dell Computer; and Ayub Khambaty, now of Chrysler; have provided many hours of research support. Additionally, Sandy Davis of GENCO, Jennie Tibben-Lembke and Deana Rogers provided valuable editorial assistance. Genet Sauer, Program Manager of the University of Nevada Center for Logistics Management, spent many hours keeping us on track and making this book a reality. She has been a great help to us and added to the fun of this project.

The amount of assistance and support from GENCO Distribution System has been phenomenal. In particular, Herb Shear, Jerry Davis, Dan Eisenhuth, Curtis Greve, Dwight "Buzzy" Wyland, and Frank Koch spent many hours in support of this project. GENCO is a third party logistics service supplier that specializes in reverse logistics. Their funding made this project possible. Their willingness to make suppliers, customers, competitors, and themselves available for interviews was extremely helpful. They are a good bunch of people who have done all they could to assist the research. We are very grateful for their support.

We would also like to thank the Reverse Logistics Executive Council for their ideas and financial support. Peter Junger of Nintendo of America, Ed Winter of Kmart Corporation, and Clay Valstad from Sears, Roebuck and Company stand out as being especially helpful.

Finally, without the support of our families, this task would have been hopeless. We have spent many hours away from home trying to learn new things.

Given the caliber of our support team, it would seem impossible that there might be errors or omissions contained in this book. However, it is possible that we omitted something in the execution of this project. Reverse logistics is a fairly new field that stands open both to much study of current practice and potential improvements. If there are any inadvertent mistakes or omissions, they are completely our responsibility.

Dale S. Rogers
Ronald S. Tibben-Lembke
Reno, Nevada
August 1998

Preface

The purpose of this book is twofold: to present an overview and introduction to reverse logistics, and to provide insights on how to manage reverse logistics well.

Reverse logistics is a new and emerging area, and as such, only a limited amount of information has been published to date. When possible, we have tried to present additional sources of information for the interested reader. However, in some chapters, such as Chapter 3 on Secondary Markets, no written information exists. When documentation was unavailable, information was gained through interviews, many of which were conducted on the condition of anonymity.

Chapter 1: Size and Importance of Reverse Logistics

1.1 Importance of Reverse Logistics

Research Scope

This project intends to define the state of the art in reverse logistics, and to determine trends and best reverse logistics practices. Part of the research charter was to determine the extent of reverse logistics activity in the United States. Most of the literature examined in preparation for this research emphasized the "green" or environmental aspects of reverse logistics. In this project, green issues are discussed, but the primary focus is on economic and supply chain issues relating to reverse logistics. The objective was to determine current practices, examine those practices, and develop information surrounding trends in reverse logistics practices.

To accomplish this task, the research team interviewed over 150 managers that have responsibility for reverse logistics. Visits were made to firms to examine, firsthand, reverse logistics processes. Also, a questionnaire was developed and mailed to 1,200 reverse logistics managers. There were 147 undeliverable questionnaires. From among the 1,053 that reached their destinations, 311 usable questionnaires were returned for a 29.53 percent response rate. A copy of the questionnaire is included in Appendix A.

What is Reverse Logistics?

Logistics is defined by The Council of Logistics Management as:

> *The process of planning, implementing, and controlling the efficient, cost effective flow of raw materials, in-process inventory, finished goods and related information from the point of origin to the point of consumption for the purpose of conforming to customer requirements.*

Reverse logistics includes all of the activities that are mentioned in the definition above. The difference is that reverse logistics encompasses all of these activities as they operate in reverse. Therefore, reverse logistics is:

> *The process of planning, implementing, and controlling the efficient, cost effective flow of raw materials, in-process inventory, finished goods and related information from the point of consumption to the point of origin for the purpose of recapturing value or proper disposal.*

More precisely, reverse logistics is the process of moving goods from their typical final destination for the purpose of capturing value, or proper disposal.

Remanufacturing and refurbishing activities also may be included in the definition of reverse logistics. Reverse logistics is more than reusing containers and recycling packaging materials. Redesigning packaging to use less material, or reducing the energy and pollution from transportation are important activities, but they might be better placed in the realm of "green" logistics. If no goods or materials are being sent "backward," the activity probably is not a reverse logistics activity.

Reverse logistics also includes processing returned merchandise due to damage, seasonal inventory, restock, salvage, recalls, and excess inventory. It also includes recycling programs, hazardous material programs, obsolete equipment disposition, and asset recovery.

Respondent Base

Companies included in this research are manufacturers, wholesalers, retailers, and service firms. In some cases, a firm may occupy more than one supply chain position. For example, many of the manufacturers are also retailers and wholesalers. The supply chain position of the research respondents is depicted in Table 1.1.

Table 1.1
Supply Chain Position

Supply Chain Position	Percentage of Respondents
Manufacturer	64.0%
Wholesaler	29.9%
Retailer	28.9%
Service Provider	9.0%

Most of the firms included in the research are very large companies. As is depicted in Figure 1.1 below, nearly half of the firms have annual sales of $1 billion or larger.

Figure 1.1
Size of Research Respondents

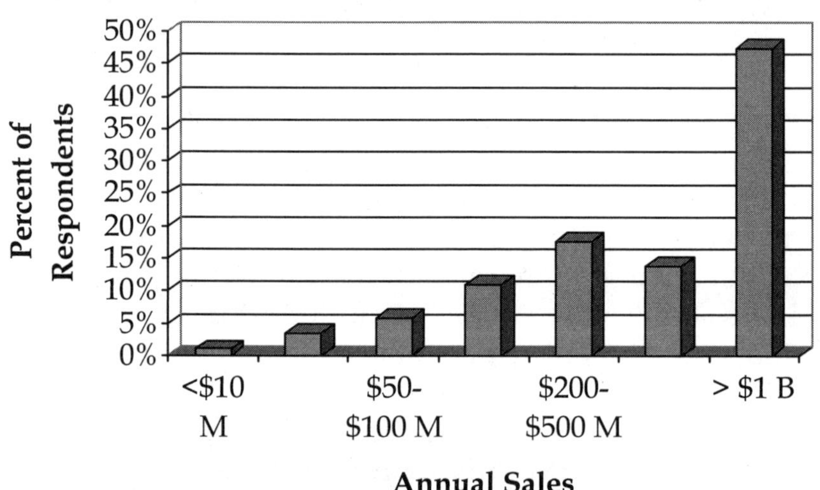

Interest in Reverse Logistics

Awareness of the art and science of logistics continues to increase. Additionally, great interest in reverse logistics has been piqued. Many companies that previously did not devote much time or energy to the management and understanding of reverse logistics, have begun to pay attention. These firms are benchmarking return operations with best-in-class operators. Some firms are even becoming

ISO certified on their return processes. Third parties specializing in returns have seen a great increase in the demand for their services.

In addition to this research project, several other academic endeavors focusing on the reverse flow of product are in process. Leading-edge companies are recognizing the strategic value of having a reverse logistics management system in place to keep goods on the retail shelf and in the warehouse fresh and in demand.

Size of Reverse Logistics

A conservative estimate is that reverse logistics accounts for a significant portion of U.S. logistics costs. Logistics costs are estimated to account for approximately 10.7 percent of the U.S. economy.[1] However, the exact amount of reverse logistics activity is difficult to determine because most companies do not know how large these are. Of the firms included in this research, reverse logistics costs accounted for approximately four percent of their total logistics costs. Applying this mean percentage to Gross Domestic Product (GDP), reverse logistics costs are estimated to be approximately a half percent of the total U.S. GDP. Delaney estimates that logistics costs accounted for $862 billion in 1997. The estimate of this research, based on the respondent sample, is that reverse logistics costs amounted to approximately $35 billion in 1997. The magnitude and impact of reverse logistics varies by industry and channel position. It also varies depending on the firm's channel choice. However, it is clear that the overall amount of reverse logistics activities in the economy is large and still growing.

Within specific industries, reverse logistics activities can be critical for the firm. Generally, in firms where the value of the product is largest, or where the return rate is greatest, much more effort has been spent in improving return processes. The auto parts industry is a good example. The remanufactured auto parts market is estimated (by the Auto Parts Remanufacturers Association) to be $36 billion.[2] For example, 90 to 95 percent of all starters and alternators sold for replacement are remanufactured. By one conservative estimate, there are currently 12,000 automobile dismantlers and remanufacturers operating in the United States.

Rebuilding and remanufacturing conserve a considerable amount of resources. According to the APRA, about 50 percent of the original starter is recovered in the rebuilding process. This may result in saving several million gallons of crude oil, steel, and other metals. APRA estimates that raw materials saved by remanufacturing worldwide would fill 155,000 railroad cars annually. That many rail cars would make a train over 1,100 miles long.

Return Percentages
The reverse logistics process can be broken into two general areas, depending on whether the reverse flow consists primarily of products, or primarily of packaging. For product returns, a high percentage is represented by customer returns. Overall customer returns are estimated to be approximately six percent across all retailers. Return

percentages for selected industries are shown in Table 1.2. In each case, return percentages were established by several different firms.

Table 1.2
Sample Return Percentages

Industry	Percent
Magazine Publishing	50%
Book Publishers	20-30%
Book Distributors	10-20%
Greeting Cards	20-30%
Catalog Retailers	18-35%
Electronic Distributors	10-12%
Computer Manufacturers	10-20%
CD-ROMs	18-25%
Printers	4-8%
Mail Order Computer Manufacturers	2-5%
Mass Merchandisers	4-15%
Auto Industry (Parts)	4-6%
Consumer Electronics	4-5%
Household Chemicals	2-3%

Clearly, return rates vary significantly by industry. For many industries, learning to manage the reverse flow is of prime importance.

Direct Retailers

Comparatively, direct or catalog companies have higher return rates than most other retail channels. It is not unusual for a direct retailer to have return rates above 35 percent. The mean level is approximately 25 percent. These catalog firms have had to improve their management of the return process. An exception to this is build-to-order, direct computer manufacturers that have lower rates of return than computer manufacturers that sell through traditional retail channels.

Most catalog firms have developed returns programs internally. They utilize their reverse logistics capabilities strategically. As the old saying goes, necessity is the mother of invention. Because return rates for many of the catalog retailers have traditionally been high, a reduction in both the number of returns and the cost of those returns was needed.

One particularly good example of skillful reverse logistics management is the J.C. Penney Catalog Division. They struggled for many years with high rates of return. Their catalog division operated independently from their retail store division. By thinking about the profitability of the whole corporation, and laying aside some difficult accounting practices, they have been able to develop a system that rewards the retail store managers for working to reduce expensive returns. When consumers decide to return a catalog purchase, they bring it back to the nearest store. The store managers are incented to disposition the item through the retail store. If the item is not sold in the store, then it is sent back to the catalog distribution center. In

Chapter 2, the difficulties of running one distribution center both forward and backward are discussed. J.C. Penney has been able to efficiently marry forward and backward distribution, primarily because reverse logistics is a priority for catalog distribution.

1.2 Reverse Logistics Activities

Typical reverse logistics activities would be the processes a company uses to collect used, damaged, unwanted (stock balancing returns), or outdated products, as well as packaging and shipping materials from the end-user or the reseller.

Once a product has been returned to a company, the firm has many disposal options from which to choose. Some of these activities are summarized in Table 1.3. If the product can be returned to the supplier for a full refund, the firm may choose this option first. If the product has not been used, it may be resold to a different customer, or it may be sold through an outlet store. If it is not of sufficient quality to be sold through either of these options, it may be sold to a salvage company that will export the product to a foreign market.

If the product cannot be sold "as is," or if the firm can significantly increase the selling price by reconditioning, refurbishing or remanufacturing the product, the firm may perform these activities before selling the product. If the firm does not perform these activities in-house, a third party firm may be contracted, or the product can be sold outright to a reconditioning/remanufacturing/refurbishing firm.

Table 1.3
Common Reverse Logistics Activities

Material	Reverse Logistics Activities
Products	Return to Supplier
	Resell
	Sell via Outlet
	Salvage
	Recondition
	Refurbish
	Remanufacture
	Reclaim Materials
	Recycle
	Landfill
Packaging	Reuse
	Refurbish
	Reclaim Materials
	Recycle
	Salvage

After performing these activities, the product may be sold as a reconditioned or remanufactured product, but not as new. If the product cannot be reconditioned in any way, because of its poor condition, legal implications, or environmental restrictions, the firm will try to dispose of the product for the least cost. Any valuable materials that can be reclaimed will be reclaimed, and any other recyclable materials will be removed before the remainder is finally sent to a landfill.

Generally, packaging materials returned to a firm will be reused. Clearly, reusable totes and pallets will be used many times before disposal. Often, damaged totes and pallets can be refurbished and returned to use. This work may be done in-house, or using companies whose sole mission is to fix broken pallets and refurbish packaging. Once repairs can no longer be made, the reusable transport packaging must be disposed of. However, before it is sent to a landfill, all salvageable materials will be reclaimed.

European firms are required by law to take back transport packaging used for their products. To reduce costs, firms attempt to reuse as much of these materials as possible, and reclaim the materials when they can no longer be reused.

Reverse Flow of Goods

The activities shown in Table 1.3 are the types that are generally considered the core of reverse logistics processes. Each of these activities gives rise to some interesting questions, many of which will be addressed in this research. However, from a logistics perspective, the larger issue common to all of these activities is how the firm should effectively and efficiently get the products from where they are not wanted to where they can be processed, reused, and salvaged. Also, the firm must determine the "disposition" of each product. That is, for each product, the firm must decide the final destination for products inserted into the reverse logistics flow.

Classifying Reverse Logistics Activities

Clearly, reverse logistics can include a wide variety of activities. These activities can be divided as follows: whether the goods in the reverse flow are coming from the end user or from another member of the distribution channel such as a retailer or distribution center; and whether the material in the reverse flow is a product or a packaging material. These two factors help to provide a basic framework for characterizing reverse logistics activities, although other important classification factors exist. Regardless of their final destination, all products in the reverse flow must be collected and sorted before being sent on to their next destinations. Where products are inserted into the reverse flow is a prime determinant in the resulting reverse logistics system.

In Table 1.4, a number of reasons for products in the reverse flow have been placed within the context of this framework. If a product enters the reverse logistics flow from a customer, it may be a defective product, or, the consumer may have claimed it was defective in order to be able to return it. The consumer may believe it to be defective even though it is really in perfect order. This category of returns is called "non-defective defectives."

If the product has not yet reached the end of its useful life, the consumer may have returned the product for service, or due to a manufacturer recall. If the product has reached the end of its useful life, the customer may, in some cases, return the product to the manufacturer so the manufacturer can dispose of the product properly, or reclaim materials.

If a supply chain partner returns a product, it is because the firm has excess product due to an over-ordered marketing promotion, or because the product failed to sell as well as desired. Also, the product may have come to the end of its life, or to the end of its regular selling season. Finally, the product may have been damaged in transit.

Table 1.4
Characterization of Items in Reverse Flow,
by Type and Origin

Source of Reverse Flow

	Supply Chain Partners	End Users
Products	Stock Balancing Returns Marketing Returns End of Life/Season Transit Damage	Defective/Unwanted Products Warranty Returns Recalls Environmental Disposal Issues
Packaging	Reusable Totes Multi-Trip Packaging Disposal Requirements	Reuse Recycling Disposal Restrictions

Given the relatively limited usage of reusable packaging in the U.S., it is reasonable to say that the majority of reverse logistics activities are related to the products only, and not to packaging. There are exceptions to this perception. A number of domestic firms are beginning to use reusable containers—such as plastic totes and knockdown cages. However, as will be described in Chapter 5, European manufacturers are required to take back the packaging for that item. In such an environment, packaging and related materials account for a very significant amount of reverse logistics activities. As more U.S. firms establish a presence in Europe, reusable packaging will become more commonplace.

1.3 Strategic Use of Reverse Logistics

Reverse Logistics as a Strategic Weapon

When companies think about strategic variables, they are contemplating business elements that have a long-term bottom line impact. Strategic variables must be managed for the viability of the firm. They are more than just tactical or operational responses to a problem or a situation.

Not long ago, the only strategic variables a firm was likely to emphasize were business functions, such as finance or marketing. During the late 1970s and 1980s, some forward-thinking companies began to view their logistics capabilities as strategic.

Although more and more firms have begun to view their ability to take back material through the supply chain as an important capability, the majority of these firms have not yet decided to emphasize reverse logistics as a strategic variable.

There is no question that the handling of reverse logistics challenges is an essential, strategic capability. In a celebrated case a few years ago, the McNeil Laboratories division of Johnson & Johnson experienced a very serious threat when someone poisoned several people by placing cyanide inside unopened bottles of Tylenol, a Johnson & Johnson flagship product. This horrible act happened twice in the space of a few years. The second time, Johnson & Johnson was prepared with a fine-tuned reverse logistics system and immediately cleansed the channel of any possibly tainted product. Because Johnson & Johnson acted so quickly and competently, a mere three days after the crisis, McNeil Laboratories experienced an all-time record sales day. Undoubtedly, the public would not have responded so positively had Johnson & Johnson not been able to quickly and efficiently handle its recalled product through its existing system in reverse. Clearly, the Tylenol incident is an extreme example, but it illustrates how reverse logistics capabilities can be strategic, and how they can dramatically impact the firm.

Another example of how reverse logistics can be used by retailers as a strategic variable is by keeping consumer product fresh and interesting. According to Dan Eisenhuth, executive vice president for asset recovery at GENCO Distribution System, "Retailers used to liquidate to compensate for 'screw-ups.' Today they do it to stay fresh."

The most important asset a retail store has is its retail space. To maximize profit per square foot of selling space, stores have to keep the fresh goods visible. Grocery stores, with razor-thin profits of one to two percent, realized long ago that it is critical to keep only products that will sell on the shelf.   Supermarkets have to turn their inventories frequently to prevent spoilage loss, and to maximize the return on their space.   Now, non-grocery retailers have begun to adapt supermarket ideas to their own businesses.

Grocery retailers started building reclamation centers in the 1970s. These reclamation centers were places where old and non-selling product would be sent. In many instances, reclamation centers would be attached to a store. Later on, supermarket chains began shipping obsolete or bad product to one central reclamation center for processing. These reclamation centers gave birth to the concept of centralized return centers, which will be discussed in greater detail in Chapter 2.

Reverse logistics is strategically used to allow forward channel participants—such as retailers and wholesalers—to reduce the risk of buying products that may not be "hot selling" items. For example, a record company developed a program to adjust return rates for various products depending on variables such as name recognition of the individual recording artist. This program produces a win-win environment for both the producer and the retailer, not to mention the consumer, who gets a broader selection. The program gives the company the ability to develop new artist franchises. Had the record company not implemented this program, its retailers would likely be willing to only carry "sure-thing" products.

Another example of the strategic use of returns is the electronic distributor that, during a period of volatile memory chip prices, created a program to help resellers better control their inventory and balance stocks. By allowing resellers to return anything within a reasonable time frame, customers were encouraged to keep inventory low and make purchases just-in-time.

Strategic uses of reverse logistics capabilities increase the switching costs of changing suppliers. A goal of almost every business is to lock customers in so that they will not move to another supplier. There are many ways to develop linkages that make it difficult and unprofitable for customers to switch to another supplier. An important service a supplier can offer to its customers is the ability to take back unsold or defective merchandise quickly, and credit the customers in a timely manner.

If retailers do not have a strategic vision of reverse logistics today, it is likely that they will be in trouble tomorrow. Retailers in high-return categories—such as catalog, toys, and electronics—can easily go out of business if they do not have a strong reverse logistics program. Given the competitive pressure on North American retailers, bottom line contributions provided by good reverse logistics programs are important to the firms' overall profitability. For more than one mass merchandiser included in the research, the bottom line impact of good reverse logistics was large. Another large retailer found that 25 percent of the profit of the entire firm was derived from its reverse logistics improvements during its initial phase.

In this research project, the research team examined several ways that reverse logistics can be utilized in a strategic manner.   These strategic uses of reverse logistics are presented in Table 1.5 below.

Table 1.5
Strategic Role of Returns

Role	Percentage
Competitive Reasons	65.2%
Clean Channel	33.4%
Legal Disposal Issues	28.9%
Recapture Value	27.5%
Recover Assets	26.5%
Protect Margin	18.4%

Competitive Reasons

Research respondents said they initiated reverse logistics as a strategic variable for competitive reasons. Most retailers and manufacturers have liberalized their return policies over the last few years due of competitive pressures. While the trend toward liberalization of return policies has begun to shift a little, firms still believe that a satisfied customer is their most important asset. Part of satisfying customers involves taking back their unwanted products or products that the customers believe do not meet needs.

Generally, customers who believe that an item does not meet their needs will return it, regardless of whether it functions properly or not. In an interesting example of this behavior, one retailer recently reported the return of two ouija boards. Ouija boards are childrens' toys that, supposedly, allow contact with the spirit world. On one ouija board there was a note describing that it did not work because "...no matter how hard we tried, we could not get any good answers from the 'other side'..." The other ouija board returner said that the reason for return was: too many spirits responded to the ouija board session, and things became too scary. In both cases, the consumers were allowed to return these "defective" products.

These competitive pressures appear to be, in large part, cultural. North American consumers and businesses are much quicker to return goods than those in most other countries. In fact, in many other countries, returns are never allowed. Some of the international managers and academics interviewed in the course of this research believed that if liberal returns were ever allowed in their country, both businesses and consumers would abuse them. However, it is clear that in some countries, business return models are moving closer to North American models. It is likely that over the next few years international firms will feel strong pressure to liberalize their return policies, and improve their reverse logistics capabilities.

Before beginning the quantitative phase of the research, it was believed that retailers had started to move away from liberal return policies that became omnipresent during the 1970s and 1980s. However, that has not been found to be the case. Respondents to this research still believe, overall, that their firms' return policies are still fairly liberal. This response is depicted in Figure 1.2 below. Respondents were asked to evaluate their returns policy on a 1 to 7 scale where 1=very conservative and 7=very liberal.

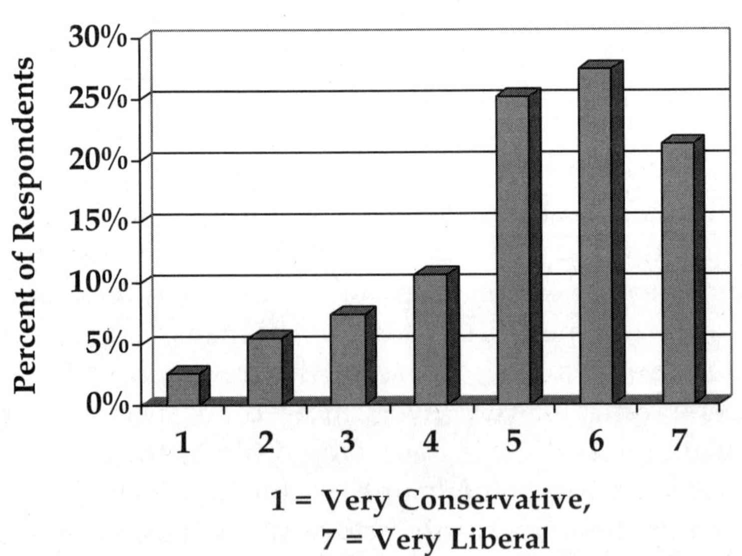

Figure 1.2
Return Policy Distribution

Return Policy Changes

Some firms have begun to take a more aggressive stance with customers, and have attempted to reduce the number of returns.   Because of customer service pressures, it is difficult to make a preemptive step, if other firms operating in the same industry have liberal return policies.   If one player in the industry has a liberal return policy, it is difficult for other firms in that industry to tighten their return policies.

Some retailers are beginning to rethink liberal return policies, and balance their value as a marketing tool against the cost of those policies.   Return policies are tightening,[3] as retailers look for ways to analyze the returns process, and to recapture dollars that were previously written on the expense side of the ledger.

One reason for a generous return policy is that it leads to improved risk sharing between sellers and consumers.   In some channels, consumers can return anything to the retailers, the retailers and wholesalers have liberal return arrangements with manufacturers, and manufacturers end up taking responsibility for the entire product life cycle. These liberal return policies occasionally turn into "return abuse" policies, where the manufacturers end up taking an inordinate amount of risk.

It is interesting to note that, overall, the research respondents do not believe that their firms' policies have changed much. In the light of celebrated examples, such as the case of an electronics retailer that began charging customers a restocking fee when returning product, the research team expected to find that return policies had begun to tighten. While tightening of return policies may develop over the next few years, as of this writing, it has not yet happened.

In Figure 1.3, perceived changes in the returns policies are presented. As can be seen from this graph, returns policies do not appear to be shifting very much.

Figure 1.3
Change in Return Policies

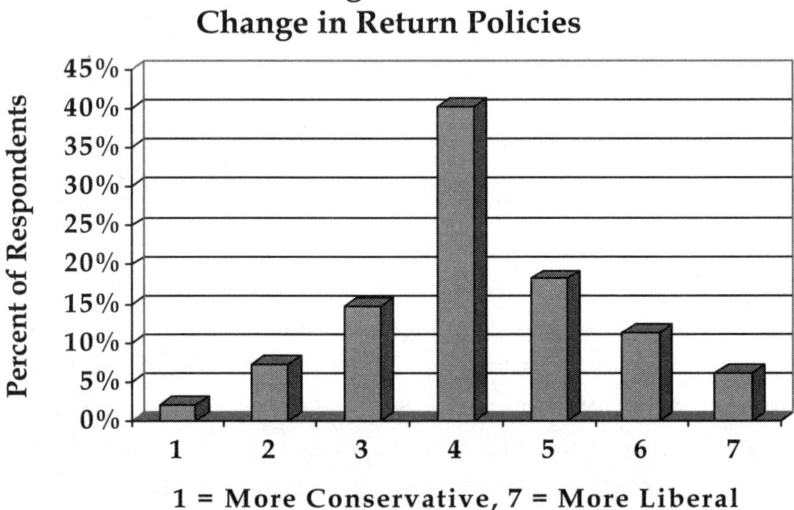

Good Corporate Citizenship

Another set of competitive reasons are those that distinguish a firm by doing well for other people. Some firms will use their reverse logistics capabilities for altruistic reasons, such as philanthropy. For example, Hanna Andersson, a $50 million direct retailer of infants and toddlers clothes, developed a program called Hannadowns.   In the Hannadowns program, customers are asked to mail back their childrens' gently worn Hanna Andersson clothes. The company then will give those customers 20 percent off the purchase price of new Hanna Andersson clothes. For Hanna Andersson, this program has been very successful. In 1996, 133,000 garments and accessories were returned.   These returns were then distributed to schools, homeless shelters, and other charities.[4]

In a second example, a shoe manufacturer and retailer, Kenneth Cole Productions, encourages consumers to return old shoes to Kenneth Cole stores during the month of February. In return for bringing in an old pair of shoes, the customer receives a 20 percent discount on a new pair of Kenneth Cole shoes.

The Kenneth Cole shoe donation program has been very successful in providing shoes to those in need. In Figure 1.4 below, an advertisement for the program is depicted. (The ad ran in February, 1998.)

Nike also encourages consumers to bring their used shoes back to the store where they were purchased. These shoes are shipped back to Nike, where they are shredded and made into basketball courts and running tracks. Instead of giving consumer discounts, like Andersson or Kenneth Cole, Nike donates the material to make basketball courts, and donates funds to help build and maintain those courts. Managing these unnecessary reverse flows is costly.

Figure 1.4
Kenneth Cole Productions
Shoe Return Advertisement

However, these activities enhance the value of the brand and are a marketing incentive to purchase their products.

In each of these examples, firms are utilizing reverse logistics strategically. They are acting as good corporate citizens, by contributing to the good of the community and assisting people who are probably less fortunate than their typical customers. While these policies may not be the reason all customers purchase their products, they are considered a marketing incentive. It is using reverse logistics to not just be environmentally friendly, but to incent customers at a real cost to their businesses.

Clean Channel

Reverse logistics competencies are also used to clean out customer inventories, so that those same customers can purchase more new goods. Auto companies have fairly liberal return policies in place, and a large reverse logistics network which allows them to bring back parts and components from their dealers. These parts are often remanufactured, so that value is reclaimed. If new parts held by the dealer are not selling well, the auto companies will give the dealers a generous return allowance, so that they can buy new parts that they really need, and therefore, service the ultimate consumer better. Most auto dealers, and many dealers in other industries, are family-based businesses with limited supplies of capital to invest in inventories. They often have less than state of the art inventory management capabilities. It is in the best interest of parts suppliers to clean out their inventories, reduce credit-line constraints, and improve customer satisfaction.

Protect Margin

Nearly 20 percent of the firms included in the research use their reverse logistics capabilities to protect their margins. This strategic usage of reverse logistics is closely related to cleaning out the channel. Firms cleanse their inventories and the inventories of their customers and their customers' customers utilizing reverse logistics processes. Some firms are proactive in their management of downstream inventory, as opposed to merely being reactive. These firms have programs in place that maximize inventory freshness. Fresher inventories can demand better prices, which in turn, protects margin.

Legal Disposal Issues

Another set of reasons named as being strategic deals with legal disposal issues. Over 25 percent of the respondents said that legal disposal issues are a major concern. As landfill fees increase, and options for disposal of hazardous material decrease, legally disposing of non-salvageable materials becomes more difficult. Firms have to think carefully about these issues. One company included in this research had previously managed hazardous waste carelessly, and experienced trouble with the Environmental Protection Agency. The result of this conflict was the primary determinant in the configuration of its manufacturing and distribution systems. This firm now wants to make sure that anything that comes out of its facilities is disposed of properly.

Recapture Value and Recover Assets

Over 25 percent of the firms included in the research said that recapturing value and recovering assets were strategic. Firms that have recently begun asset recovery programs found that a surprisingly large portion of their bottom-line profits is derived from asset recovery programs. These programs add profit derived from materials that were previously discarded, which makes them essentially free.

Conclusions

While many companies have yet to recognize the strategic potential of efficient reverse logistics, it is clear that the tide is beginning to turn.   There is more interest in reverse logistics now than ever before. Firms are beginning to make serious investments in their reverse logistics systems and organizations.   One clear indication of the strategic importance of a business element is the amount of money spent on managing that element.

Given the volume of returned products experienced in some industries, it is not surprising that the firms in those industries consider returns a strategic and core competency. It appears likely that companies in industries that generally do not place much value on good reverse logistics practices, will, over the next few years, find that making investments in their return systems will enhance their profitability. It is clear that for many firms, excellent reverse logistics practices add considerably to their bottom line.

1.4 Reverse Logistics Challenges

Retailer   Manufacturer Conflict

One of the difficulties in managing returns is the difference in the objectives of manufacturers and retailers.   The distance between them on many issues can make the difference seem like a chasm, as shown in Figure 1.5.

Figure 1.5
The Chasm Between Manufacturer and Retailer

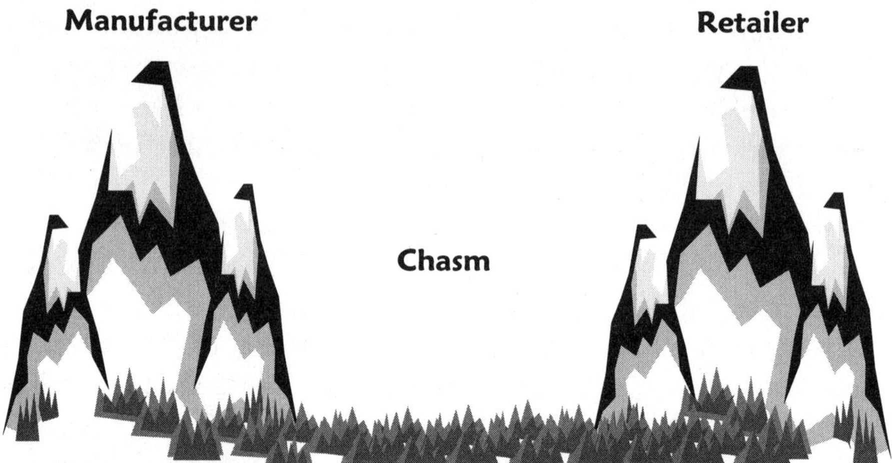

Source: Clay Valstad, Sears, Roebuck and Co.

Whenever a retailer wants to return an item, the retailer and the manufacturer may disagree on any one of the following:

- Condition of the item
- Value of the item
- Timeliness of response

Often from the retailer's perspective, every product was sent back in pristine condition, and any damages must have occurred in transit or must be manufacturing defects. The manufacturer may suspect the retailer of abusing return privileges because of poor planning, or of returning product damaged by the retailer. Once the condition of the item is agreed upon, the value that the retailer should receive must be determined. The retailer may claim full credit, and the manufacturer may have a dozen reasons why it should not receive full credit. These issues can be difficult to sort out. After they have all been decided, the refund never comes quickly enough to suit the retailer.

Retailer returns to the supplier are a method of reducing inventories near the end of a quarter. Retailers may suddenly move material back to the supplier, or at least notify the supplier that they are going to do so, and negotiate the details later.

For similar reasons, manufacturers can be slow to recognize returns as a subtraction from sales. They may want to delay returns until a later accounting period, or, they may not want to credit the returned items at their full price.

Sometimes the retailer simply deducts the cost of the items from an invoice. Often, that invoice is not the same one for the goods being returned.

In the end, both parties need to realize that they have to develop a working partnership to derive mutual benefit. Obviously, neither can live without the other; they need to work together to reduce the number of returns coming back and speed up the processing of those that do come back. Inefficiencies that lengthen the time for processing returns cause harm to both firms.

Problem Return Symptoms
Dr. Richard Dawe of the Fritz Institute of International Logistics identified six symptoms of problem returns.[5] Those six symptoms are depicted in Table 1.6 below.

Table 1.6
Problem Return Symptoms

<u>Symptoms</u>
- Returns arriving faster than processing or disposal
- Large amount of returns inventory held in the warehouse
- Unidentified or unauthorized returns
- Lengthy processing cycle times
- Unknown total cost of the returns process
- Customers have lost confidence in the repair activity

If a large amount of returns inventory is being held in the warehouse, clearly there is a problem with the way the firm is handling returns. If a large number of unauthorized or unidentified items are being discovered, again, there must be a significant problem with the return process.

Piles of unprocessed returns are easy to observe. Unfortunately, some of these other symptoms Dr. Dawe identified are not as easily observed. One of the findings of this research is that shortening returns processing time is important for handling returns well. If firms do not monitor the length of their processing cycle times, they have no way to determine how well they are doing in this area. One of the biggest challenges facing firms dealing with reverse logistics is a lack of information about the process. Again and again, we have seen companies that do not have any formalized systems for monitoring their reverse logistics activities. As the old saying goes, if you aren't measuring it, you aren't managing it.

Cause and Effect

Poor data collection leads to uncertainty about return causes. In the long run, the most valuable outcome of sound reverse logistics management is the accumulation of data. Improving the return process and efficiently handling the returned products decreases costs. However, being able to see defective products and to track return issues by reason codes can be more useful than simply improving return handling efficiencies. In forward distribution, it is more important to be able to manage information effectively than to mange inventory. Generally, those firms that manage information well also manage their inventories effectively. Those that do not manage well the data surrounding their logistics processes, do not generally manage their inventories effectively. This same rule applies to reverse logistics as well.

Reactive Response

Over the last few years, many companies have practiced reverse logistics primarily because of government regulation or pressure from environmental agencies; not for economic gain. For most of these companies, reverse logistics has not been as strongly emphasized as other business activities. For many firms, it has not been possible to justify a large investment in improving reverse logistics systems and capabilities because generally, not enough analysis is completed. Like the captain of the Titanic, whose disregard of iceberg warnings brought so much devastation, executives usually disregard reverse logistics issues.

1.5   Barriers to Good Reverse Logistics

As we continued to examine the firms included in this research project, it was clear that for many companies, it is difficult to successfully execute reverse logistics because of very real internal and external barriers. We asked the 300 research respondents about what kinds of issues cause them difficulty in completing their reverse logistics mission. These answers were grouped around the following categories: importance of reverse logistics relative to other issues, company policies, lack of systems, competitive issues, management inattention, financial resources, personnel resources, and legal issues. The responses are listed below in Table 1.7.

Very few of the firms interviewed manage their reverse logistics costs at the operational level. Since successfully completing the reverse logistics mission is clearly a problem for many firms, it is obvious that numerous barriers to good reverse logistics exist. According to the research respondents, the relative unimportance of reverse logistics issues (39.2 percent) is the largest barrier to good reverse logistics management. These companies said that reverse logistics was just not a priority. Some firms included in the research mentioned that they have difficulty cost-justifying a reverse logistics system. As one executive said, "after all, it is junk. You can't expect my VP to want to invest in junk." While it is not necessarily junk, it is often viewed as such and therefore is not worthy of much investment.

Table 1.7
Barriers to Reverse Logistics

Barrier	Percentage
Importance of reverse logistics relative to other issues	39.2%
Company policies	35.0%
Lack of systems	34.3%
Competitive issues	33.7%
Management inattention	26.8%
Financial resources	19.0%
Personnel resources	19.0%
Legal issues	14.1%

For many of the firms examined, this attitude is changing. For example, in the book industry, reverse logistics has traditionally not been recognized as a significant factor. Recently, high returns have pushed many publishers to operate in the red. It is clear that, in the long run, these publishers cannot continue to overlook the necessity of good reverse logistics management. As discussed in more detail in Chapter 6, returns are now considered to be extremely important in the book industry.

The second largest number of respondents mentioned restrictive company policies (35.0 percent). This response may be related to management inattention and the lack of importance of reverse logistics. It also is related to corporate strategy for handling returns and non-salable items. Because companies do not want to see their "junk" cannibalizing their first quality or "A" channel, they often develop policies that make it very difficult to handle returns efficiently, or to recover much secondary value from those returns. One trend that is interesting, however, is that the pendulum currently appears to be swinging toward eliminating difficult policies and attempting to handle returns effectively, in order to recover value from what can be a very valuable resource.

Lack of systems is another serious problem for 34 percent of the respondent base. In the course of this research project, very few good reverse logistics management systems were found.

Competitive issues (33.7 percent) and management inattention (26.8 percent) also hamper reverse logistics efforts. Financial and personnel issues were cited as barriers by 19 percent of those surveyed. This number was lower than expected although it is not insignificant. For most firms, executive attention and policies are much greater problems than adequate access to resources.

The problem that appears to have the smallest impact on reverse logistics managers is legal issues. This finding is contrary to what was expected. The conventional wisdom has been that over the last few years, most companies have practiced reverse logistics primarily because of government regulation or pressure from environmental agencies, and not for economic gain. While this may be true, legal issues do not appear to be a major problem for most of the firms included in our research.

Chapter 2: Managing Returns

There are many different kinds of reverse logistics activities. As discussed in Chapter 1, much of the focus of this research project was directed at examining the return flow of product from a retailer back through the supply chain toward its original source, or to some other disposition.

The management of this flow of materials is the focus of Chapter 2. As it will become clear, the diverse modalities for handling returns utilized by the research respondents can either positively or negatively impact a company's bottom line. What follows is a detailed examination of those factors defined by the research team as key reverse logistics management elements.

Table 2.1
Key Reverse Logistics Management Elements

- Gatekeeping
- Compacting Disposition Cycle Time
- Reverse Logistics Information Systems
- Centralized Return Centers
- Zero Returns
- Remanufacture and Refurbishment
- Asset Recovery
- Negotiation
- Financial Management
- Outsourcing

2.1 Improve Return Gatekeeping

For years, retailers and manufacturers have focused solely on massaging profitability into and out of the inventory management process—but only from a forward distribution perspective. Our research shows that the time has come to give similarly focused attention to the reverse logistics management function—and every company has one. Point of entry into the reverse logistics pipeline—or "gatekeeping," as we call it— deserves much more attention. Gatekeeping is the screening of defective and unwarranted returned merchandise at the entry point into the reverse logistics process. Good gatekeeping is the first critical factor in making the entire reverse flow manageable and profitable.

Successful companies have satisfied customers. Retail success stories, such as that of L.L. Bean, can be attributed, in large part, to excellent customer service through customer-oriented marketing, which often includes a liberal return policy. L.L. Bean is famous for being willing to accept worn-out apparel and giving the customer full credit. L.L. Bean accepts all of the risk associated with purchasing one of their products. This policy is a significant marketing incentive. During the late 1980s and early 1990s, many companies studied L.L. Bean as an example of excellent customer service. The concept of absorbing the risk that a product might be faulty, damaged, or simply unwanted, attracts customers, increases sales, and at the same time, causes major problems for retailers.

While liberal return policies draw customers, they can also encourage consumer abuse. For example, at one GENCO retail centralized return center visited by the research team, some of the items brought into the center were not even sold by the retailer to which they were returned. Retail store personnel should have never accepted those items as returns. However, without good systems in place and well trained personnel at store level, this kind of abuse occurs more often than retailers would like to admit.

In the book industry, publishers allow bookstores to return any product for credit. Often, the return rates on a specific book actually determine its profitability. Conversely, book distributors, who are the largest customers of publishers, only take back a certain percentage of the books that they sell to bookstores. Book retailers are painfully aware of the policy mismatch between publishers and distributors. Once the stores return their quota of allowable returns to their distributors, they begin sending the remainder of their returns back to the publisher — even though they bought the books from the distributor and not directly from the publisher. In some cases, book retailers do not even try to ship the product back to their distributors because of tighter distributor policies.

Using this strategy forces the publisher to incur the lion's share of the cost for book returns. Since the publisher probably sold the books originally to the distributor at a lower price than the direct price to the retail bookseller, the publisher's profits are diminished, while the distributor avoids incurring the expense of handling the return. This system hardly seems efficient or fair.

During the course of our research, several retailers voiced concern and consternation over the difficulty in screening defective, and unwarranted returned merchandise at the store level. Store-level clerks and front-line personnel are often unwilling or unable to gatekeep the returns process. Retailers need to do better training of the sales associates. They can also develop systems to take the decisions out of the hands of the associate.

Nintendo, the electronic game manufacturer, has developed a particularly innovative gatekeeping system. They rebate retailers $0.50 if they register the game player sold to the consumer at the point of sale. Nintendo and the retailer can then can determine if the product is in warranty, and also if it is being returned inside the allowed time window. They developed special packaging with a window that allows the serial number to be scanned by the retailer's point-of-sale scanner. This information updates a database that a retailer can access when the customer brings back a Nintendo machine.

The impact from this new system on their bottom line was substantial. After implementing this system, Nintendo experienced more than an 80 percent drop in return rates—to less than 2 percent of sales. However, for most manufacturers and retailers, it is too expensive to register at the point of sale $20 items. In most systems, once the sales associate makes a decision about a return, that decision is usually not overturned. Systematic problems are magnified because many sales associates do not receive much training in this area.

Failure in returns gatekeeping can also create significant friction between supplier and customer firms, not to mention lost revenue. For example, the stock price of a specialty apparel manufacturer fell dramatically at the end of 1996. This drop was due to, in large part, the inability of the specialty retailer who sold the product to appropriately manage returns to the manufacturer. The retailer, a store found in most suburban shopping malls, accounts for approximately one third of the manufacturer's revenues. Here's what caused the problem.

Instead of using a centralized return processing center, which significantly expedites the reverse logistics pipeline, the retailer accumulated store returns and sent them back to the manufacturer in infrequent, large batches. This practice, coupled with a breakdown in manufacturer-retailer communication channels, created mountains of returned product on which the retailer only received a fraction of the original cost. Subsequently, the retailer's third quarter profits suffered, and buying volumes were reduced with the manufacturer. Needless to say, Wall Street reacted negatively. The manufacturer's stock fell to a third of its high point for the year. As of this writing, both firms have been seriously wounded. These are wounds that could have been avoided if the gatekeeping function of their return process had been a priority—not a postscript.

2.2 Compact Disposition Cycle Time

Another critical element to successful reverse logistics management is having short disposition cycle times.

The companies that are best at managing their reverse logistics processes are adept at gatekeeping, as described above.   These firms are also able to reduce cycle times related to return product decisions, movement, and processing. One executive described difficulties in managing the return process and said, "You know, this stuff isn't like fine wine. It doesn't get any better with age."

While most returned product does not age well, it is clear that many firms have not discovered how to avert a lengthy aging process on their returns.   For many of the firms studied, returns are exception-driven processes.   Often, when material often comes back in to a distribution center, it is not clear whether the items are: defective, can be reused or refurbished, or need to be sent to a landfill. The challenge of running a distribution system in forward is difficult; it is harder still for companies to allocate resources to manage the system in reverse.

Part of the difficulty that firms have in compacting disposition cycle time is that there does not seem to be much reward for taking responsibility and making a timely decision as to how product should be dispositioned. Employees have difficulty making decisions when the decision rules are not clearly stated and exceptions are often made. It is easier to pass the product back to the previous stage in the channel, because that reduces both personal and company risk.

2.3 Reverse Logistics Information Systems

One of the most serious problems that firms face in the execution of a reverse logistics operation is the dearth of good information systems. Very few firms have successfully automated the information surrounding the return process. Based on the response of firms included in the research, reverse logisticians seem to feel that nearly zero good reverse logistics management information systems are commercially available. Because information systems resources are usually stretched to their limit, those resources are usually not available for reverse logistics applications. An information systems department queue for building applications not determined to be core processes is often greater than one year. Some information systems departments have queues that stretch out beyond two years. Given this difficulty, reverse logistics applications typically are not a priority for information systems departments.

To work well, a reverse logistics information system has to be flexible. In addition to the problems described above, automation of those processes is difficult because reverse logistics processes have so many exceptions. Reverse logistics is typically a boundary-spanning process between firms or business units of the same company. Developing systems that have to work across boundaries adds additional complexity to the problem.

For the retailer, a system that tracks returns at store level is desirable. The system should create a database at the store level so that the retailer can begin tracking returned product and follow it all the way back through the pipeline.

One of the best firms included in this research developed a very simple system to assist in the compacting of the disposition cycle times. In addition to an investment in computer systems, they have designed manual systems to improve returns processing. They use a three-color system. A store employee receives instructions about the returned good from decision rules built in to the point-of-sale terminal at the service desk. The point-of-sale terminal retrieves the return policy for that particular item. The store clerk places a yellow sticker on the item if it is to be returned to the vendor. A green sticker means that the item is to be placed on the salvage pallet. If the system indicates "red," the item is an exception article and has to be researched. This particular firm tries to keep the number of red stickers to a minimum. Because the disposition decision is made by the system and does not rely on individual judgements for most returns, disposition cycle time is dramatically reduced.

Additionally, because of their systems, this firm has the benefit of tracking returns, and measuring cycle times and vendor performance. This firm's buyers have much better information in their hands when they talk to suppliers and negotiate allowances. Also, the stores can see if the consumers are committing "return abuse," and are trying to take advantage of the store. These benefits have been realized because this firm has recognized the bottom-line impact of reverse logistics and assigned its resources to work on reverse logistics systems problems.

Returns Transaction Processing

In a truly integrated supply chain, everyone in the supply chain can track product as it moves forward through the channel. While there are very few supply chains that really function this well, there are virtually none that work in reverse. Most firms cannot track returns within their own organization, much less somewhere outside of their firm.

Retailer

In a returns processing system that may reside at a centralized return center, several transactions can occur. A good system might include the following steps. The first transaction will likely be financial, where an inventory category will be updated. A chargeback to reconcile with the vendor, or something similar, will occur. A retailer may want to reorder first quality product from its supplier immediately. Then, routing for processing or a storage location within the processing center will be determined. A reverse warehouse management system may be required for this step.

Manufacturer

The manufacturer will generate a return authorization (RA). This is often a manual process. RAs could be generated electronically, including an automatic check to see if the return should be authorized. Next, the likely financial impact of the return could be generated. These capabilities would be very helpful in better managing returns. The next step is to automate pickup of product and an advanced shipping notification (ASN) could be cut.

After it is shipped, it is received. Currently, most manufacturers manually receive returns. Once the material is received, a database is created for reconciliation. Because most manufacturers manually receive material, this database is created slowly — if it is created at all. This sluggishness results in slowing the reconciliation and the disposition of the returns.

EDI Standards

Electronic data interchange (EDI) standards to facilitate this boundary spanning have been developed to handle returns. The 180 transaction set was developed to manage the flow of information surrounding the return process. However, few of the research respondents have implemented the 180 EDI transaction set. The majority of the respondent firms have implemented some EDI functionality. They just have not put many resources into developing EDI linkages for the return flow of goods. One executive said that: "I can get suppliers to send me ASNs all day. I just can't get anyone to tell me product is coming back to the warehouse." A complete description of the 180 transaction set is given in Appendix D.

Some of the firms interviewed voiced the opinion that eventually, the internet will replace the implementation of EDI transactions. In an application such as reverse logistics, where resources are always difficult to gather, inexpensive browser-based return interfaces may be one answer to the systems problem. In addition to being less expensive, internet-style interfaces can usually be developed more quickly than costly mainframe applications. Additionally, GENCO, IBM, HP, and other firms are testing license plates and two-dimensional bar codes to fill gaps between systems. Hardware firms such as Symbol and Telxon are developing solutions for reverse logistics applications.

A good reverse logistics system can remove functionality from the back of a retail store. One retail firm interviewed for this research project found that after they installed a reverse logistics system, they were able to reduce headcount.

A good system allows the firm to quickly obtain credit for returned product, which improves cash flow management through the reverse logistics pipeline. A company can change suppliers, liquidate the old supplier's product, and get through final resolution much more quickly than if the reverse logistics information flow is not automated.

Return Reason and Disposition Codes

Part of good returns transaction processing is understanding why the items were returned and how they should be dispositioned. Listed below in Table 2.2 are possible standardized return reason codes.

Table 2.2
Return Reason Codes

Repair / Service Codes
- Factory Repair – Return to vendor for repair
- Service / Maintenance
- Agent Order Error – Sales agent ordering error
- Customer Order Error – Ordered wrong material
- Entry Error – System processing error
- Shipping Error – Shipped wrong material
- Incomplete Shipment – Ordered items missing
- Wrong Quantity
- Duplicate Shipment
- Duplicate Customer Order
- Not Ordered
- Missing Part

Damaged / Defective
- Damaged – Cosmetic
- Dead on Arrival – Did not work
- Defective – Not working correctly

Contractual Agreements
- Stock Excess – Too much stock on hand
- Stock Adjustment – Rotation of stock
- Obsolete – Outdated

Other
- Freight Claim – Damaged during shipment
- Miscellaneous

In Table 2.3 below, potential disposition codes are presented.

Table 2.3
Disposition Codes

Disposal
- Scrap / Destroy
- Secure Disposal
- Secure Disposal (Videotaped)
- Donate to Charity
- Third Party Disposal
- Salvage
- Third Party Sale (Secondary Markets)

Repair / Modify
- Rework
- Remanufacture / Refurbish
- Modify (Configurable or Upgradable Products)
- Repair
- Return to Vendor

Other
- Use as Is
- Resale
- Exchange
- Miscellaneous

Several companies included in the research have also taken a larger, more difficult step in compacting disposition cycle times. This step is the development of a centralized return center (CRC) network. While it is not intuitively clear that establishing CRCs would reduce cycle times, in every firm studied that moved to the CRC concept, disposition times decreased. This reduction in time is most likely due to improved information systems and clearly understood procedures for handling returned material. In most cases examined, this reduction in cycle time directly and positively impacted the firm's bottom line.

2.4 Centralized Return Centers

Centralized return centers (CRCs) are processing facilities devoted to handling returns quickly and efficiently. CRCs have been utilized for many years, but in the last few years, they have become much more popular as more retailers and manufacturers have decided to devote specialized buildings and workforces to managing and processing returns.

In a centralized system, all products for the reverse logistics pipline are brought to a central facility, where they are sorted, processed, and then shipped to their next destinations. This system has the benefit of creating the largest possible volumes for each of the reverse logistics flow customers, which often leads to higher revenues for the returned items. It also allows the firm to maximize its return on the items, due, in part, to sortation specialists who develop expertise in certain areas and can consistently find the best destination for each product.

Generally, centralized return centers work in the following manner. The retail stores send product back to one or more centralized return centers. If the retailer is a large, national or international company, it is likely that it will have more than one CRC. For example, Kmart Corporation has four CRCs in its system, and Sears, Roebuck and Company has three. The CRC then accumulates the returned product for processing. Generally, the CRC will make a decision about the appropriate disposition for the product, based on guidelines set by the retailer and manufacturers.

One of the most important activities is the sortation step. During this part of the process, employees make decisions about whether the product can be resold or if it has to be scrapped. Obviously, determining the best channel for dispositioning the product is of critical importance in maximizing revenue from the products in the reverse logistics pipeline.

Based on the research interviews, centralized return centers are an important part of a reverse logistics management strategy. These centers impose order on the reverse flow. Generally, they are associated with information system improvement. To run a CRC, a firm must have some sort of reverse logistics system in place. In almost every instance, research respondents said that centralized return centers had a positive impact on the bottom line. In one case, a large company said that the combination of implementing centralized return centers, moving to an asset recovery program, and improving its reverse logistics information systems improved the corporate bottom line by 25 percent.

The amount of product that a network of CRCs processes for the large retailers can be huge. One retailer included in the research ran over $800 million of product through its network of CRCs during fiscal 1997.

CRCs also simplify in-store processes. It is often difficult to get disposition decision uniformity across a chain of stores for several reasons. The employees working the customer service desk may be not properly trained, new, or not terribly concerned about returns.

Consistency

Sending returns back to a CRC results in more consistent decisions being made about product disposition. Because processes are standardized, errors are more easily identified and avoided. The quality of returns processing generally improves as the firm moves to a centralized processing model.

Space Utilization

Retail stores generally have very limited space in the store to devote to returns. Usually, a retail store wants to devote as much space as possible to the selling floor. A retailer does not want to devote much space to hold non-selling returns.

Labor Savings

By centralizing returns processing, a retailer minimizes the labor required to complete the processing of returns. One properly trained employee at the CRC can generally do more in less time than the combined efforts of several customer service desk employees.

Transportation Costs

Many of the companies included in this research also found that their reverse logistics-related transportation costs decline due to consolidation. With a CRC model, a retailer or manufacturer can utilize "milk runs" to pick up returned goods. This way, a company can move more pallets and fewer boxes, increasing consolidation and thereby reducing freight costs.

The downside to a completely centralized system is that handling and transportation costs can increase because all products must be transported from the retail locations to the centralized facility. If a product is going to be thrown away, transporting it to a centralized facility just to throw it away increases costs, but does not increase revenues, because the product is still thrown away.

However, the cost savings, reduced disposition time, and improved revenues associated with the implementation of CRCs more than make up for transportation costs incurred if the product is to be scrapped. In many cases, products that are going to be thrown away would not be processed at the retail store, anyway.

Improved Customer Service

From the manufacturers' point of view, the centralized model can improve customer service. It can speed the reconciliation process, improve return material authorization (RMA) verification, and be part of developing important management information. Because of consolidation of returns, a manufacturer can more easily become aware of trends in returns. Also, good reverse logistics management can be a marketing strategy to keep customers loyal. Processing the transaction quickly and giving the customer credit helps to build customer loyalty. Some firms believe that their returns management processes give them a great opportunity to please the customer.

Establishing CRCs is a sign of commitment from the firm to incorporate returns management into the overall corporate strategic plan. It means that someone such as the general manager of the CRC has the job of making sure that returns are handled properly.

Compacting Disposition Time

Firms included in the research that established a CRC found that their disposition cycle times were reduced.

Centralized return centers tend to expedite the reverse logistics pipeline. Before implementing a CRC, retailers would accumulate store returns and send them back to the manufacturer in infrequent, unorganized, large batches. Because returns were not normally the first priority of the store or the distribution center, returned goods would pile up in large amounts. This inefficient handling would result in the loss of value as the returned product sat for a long period of time and was often damaged. Retailers would receive less credit from the manufacturers or distributors than their original purchase price. For products such as personal computers, this situation is catastrophic, as the product loses value everyday it sits idle.

One retailer included in the research uses the faster processing and systems linkages that are part of its CRC network to assist the corporation in managing its bottom line from one quarter to the next. From an accounting point of view, this company transfers the inventory back to the supplier once a bill of lading is cut. The director of reverse logistics often receives calls near the end of the month or end of the quarter asking him if he can, for example, "get rid of $4 million and shift it back on the vendors before the period closes."

Profit Impact
Returns have a lower impact on the profitability of those firms utilizing outsourced centralized return centers than those not using outsourced centralized return centers. As Table 2.4 shows, companies that used an in-house CRC found that reverse logistics costs reduced profitability by 4.8 percent, while those companies that used a third party to manage their CRC found profitability reduced by 3.7 percent.

Table 2.4
Impact on Profitability

Activity	In-House	Outsourced
Central Return Center	4.8%	3.7%

Visibility of Quality Problems

One of the advantages related to operating a CRC is that it becomes easier to see quality problems as product flows in from several retail stores. Several of the firms that operate CRCs told the research team that if the firm is doing a good job of gatekeeping, and has a system in place that allows it to match returned merchandise with the vendor file, the firm can more quickly see problem products and suppliers. They can then improve product quality and reduce returns.

An example of this is the experience that a major retailer had with some dehumidifiers that inexplicably began arriving at their CRC in large numbers. The CRC team called the buyer, who called the manufacturer. The manufacturer sent an engineer to examine the problem. The engineer laid the dehumidifiers out on a workbench and found a plastic liner that was melting. The melting plastic liner resulted in a faulty product. Because the CRC effectively managed information, a problem was solved that, otherwise, could have resulted in much greater expense for all of the members of the supply chain.

By tracking hundreds of return authorizations, a firm can build a data warehouse that contains return reasons. If a quality problem exists with a product, consolidation of returns will highlight those quality difficulties more quickly than if returns dribble in slowly from retail customer service desks. A consumer electronics company was seeing high return rates on personal CD players. Jogging while wearing or carrying one of these portable CD players made them skip. The returns were sent to a CRC where they were processed. Management information for a large number of these CD players was developed, which gave the firm important feedback and enabled it to solve the problem.

In another example, a firm that manufactures bread machines was experiencing a high rate of return on apparently operable machines. The problem was that the picture on the box showed a perfectly formed loaf of bread. The bread machine, however, actually produced a loaf that looked like a ball. It was not until the manufacturer could view management information that was derived from a number of returns that they understood the actual problem.

In the electronics industry, or in other industries that produce and sell "high learning products," a large number of returns are not really defective. A high learning product is one that requires users to do more than simply unwrap it and put it in their mouths, play with it, or simply turn it on. Products that require user knowledge or expertise for proper operation often come back in large numbers. The ones that actually work are "non-defective defective" products. There is really not anything wrong with them, but the consumer could not figure out how to make them work. By utilizing a CRC and seeing the return reasons in larger numbers, the manufacturer and the retailer can work together to improve the manual, develop a quickstart sheet, give an 800 phone number, or come up with another solution to reduce the non-defective defectives. For some products, the highest percentage of returns is actually non-defective defectives.

Forward and Backward
A large number of firms interviewed believes that distribution centers will not work well both forward and backward. While at first this belief did not seem logical to the research team, many distribution centers that attempt to efficiently process both forward and reverse supply chain flows have much difficulty. The problem may be related more to focus than to actual capability. If the distribution center manager has to make a choice between efficiently executing forward logistics versus reverse logistics, the distribution center manager will focus on forward distribution. In every situation, forward distribution management is the top priority of a distribution center that has new product flow as one of its responsibilities.

A few of the research respondents said that cycle time processing is negatively affected when one distribution center handles both forward and reverse shipments. In distribution centers that have a limited number of dock doors and dock space, product coming back tends to be mishandled and the processing of that material is often postponed.

This problem of utilizing a distribution center to work both forward and backward is one of the reasons that several firms are seeking out specialists. It is difficult for executives to justify the expense of constructing and staffing a large building dedicated to handling "failures"–which is exactly how returned product is often perceived inside the company.

Some firms included in the research are able to perform forward and reverse logistics operations in the same facility by carefully segregating both the flow of product and the employees. This separation allows the reverse logistics employees to focus on the return flow and not be distracted by forward distribution activities.

Accounting Issues

In a good CRC, information systems interact with accounting and other systems. In theory, stores should be able to do this well, but unfortunately, most cannot. For the most part, retailers want to perform tasks that are part of the normal retail process. A CRC assists the retailer and enables it to make faster disposition decisions about returned product.

To summarize, the benefits resulting from a centralized return center are presented in Table 2.5.

**Table 2.5
Typical Benefits from a
Centralized Return Center**

1. Simplified store procedures
2. Improved supplier relationships
3. Better returns inventory control
4. Improved inventory turns
5. Reduced administrative costs
6. Reduced store level costs
7. Reduced shrinkage
8. Refocus on retailer core competencies
9   Reduced landfill
10. Improved management information

2.5 Zero Returns

In zero return programs, the manufacturer or distributor does not permit products to come back through the return channel. Instead, they give the retailer or other downstream entity a return allowance, and develop rules and guidelines for acceptable disposition of the product. A typical return allowance in many industries is three-and-a-half to four percent of sales to the retailer.

A zero returns policy, properly executed, can result in substantially lower costs, according to the research respondents. Firms using zero returns can reduce the variability of returns costs, by pre-setting the maximum dollar amount of returned product. Stabilizing return rates using a zero returns program promotes planning and fiscal health.

Zero returns enables the firm to avoid physically accepting returns altogether, a strategy being adopted by some consumer product companies, and several electronics companies.

Interestingly, most retailers do not track the cost of returns. Instead, merchandise buyers factor the return allowance into their pricing, which ignores the cost of returns.

While zero returns release upstream channel participants from dealing with the physical portion of reverse logistics management, it does not reduce much of the physical burden placed on downstream channel participants.

Thus far, zero return programs have reportedly had mixed results. At a large consumer products company, zero returns appear to have reduced the handling costs related to returns. However, much of the product that this firm earmarks as scrap-in-the-field is, instead, showing up in alternative channels, such as "dollar" stores and flea markets. Cannibalization of the "A" or primary channel is always a concern when considering a zero returns program.

How Zero Returns Work

In a typical zero returns program, a supplier tells its customers that no product will be accepted for return, once ordered. Instead, the supplier will give the customer a discount off of the invoiced amount. Depending on the supplier, the retailer either destroys the product, or disposes of it in some other manner.

In another model being utilized in the computer industry, the retailer returns all product to a central point on open return material authorization (RMA). Usable product is paid for and shipped to a third party for refurbishment and disposition. Ineligible or unusable product is disposed of based on a predefined set of rules. In this model, the goal for the retailer is to enlist as many manufacturers as possible to participate, to enable centralized receiving, auditing, and payment processing. From the computer manufacturer's point of view, all product from the channel is returned to the refurbishing third party. The product is audited by an independent entity to determine its usability and the retailer's credit entitlement. All aspects of the RMA process and the disposition of the returned product are handled by the third party.

In this model, ineligible product becomes moot, because the third party settles claims between retailers and manufacturers. The result for the retailer may be faster reconciliation and payment; simplified, easier returns processing; and cheaper, reduced inventory awaiting return to the vendor. For the manufacturers, this model may result in a faster recovery process, better RMA administration, and a renewed focus on selling new products, which is their primary goal.

2% / 6% Problem

A notable problem with zero return policies is one that can be referred to as the "2% / 6%" problem. Because of the considerable power that large retailers have in the channel, it is hard for manufacturers to decree an appropriate returns allowance and stick to it. For example, if a manufacturer is selling product to Kmart, and sets a six percent returns cap, Kmart would agree to that cap if the returns of the manufacturer's product do not exceed six percent. In fact, Kmart would probably be very happy if its actual returns rate was two percent while it was receiving credit for six percent returns. Kmart would be able to use the additional return cap money as a rebate.

However, if the manufacturer sets a two percent return cap and the actual return rate is six percent, Kmart is not likely to consent to the manufacturer's set return cap percentage. The retailer would instead insist that the manufacturer cover the entire six percent returns. Because of the power of the large retailers, most manufacturers are not in a position to argue about the return cap percentage. This, coupled with the inability of the manufacturer to truly control the disposition of the product, means that the retailer has less risk than the manufacturer from a zero returns program. An effective zero returns programs requires that both the buyer and seller truly understand what their actual costs are.

2.6 Remanufacture and Refurbishment

Thierry, et al. (1995) defined five categories of remanufacture and refurbishment. These five categories, shown in Table 2.6, are repair, refurbishing, remanufacturing, cannibalization, and recycling. The first three categories: repair, refurbishing, and remanufacturing, involve product recondition and upgrade. These options differ with respect to the degree of improvement. Repair involves the least amount of effort to upgrade the product, and remanufacture involves the greatest.

Cannibalization is simply the recovery of a restricted set of reusable parts from used products. Recycling is the reuse of materials that were part of another product or subassembly.[1]

Table 2.6
Remanufacturing and Refurbishment Categories

1. Repair
2. Refurbishing
3. Remanufacturing
4. Cannibalization
5. Recycling

Remanufacturing and refurbishing of used product is on the rise. Even NASA spacecraft are being built with remanufactured and refurbished tools. A prime subcontractor for NASA utilizes remanufactured machine tools to produce complex spherical components for spacecraft. Remanufacturing was chosen over purchasing new equipment to generate cost savings. Customization of the machines was inevitable to adapt them to desired computer numerical control and spindle combination standards. In this particular case, older machines were built with heavier castings. Because the mass of these older machines is greater than newer models, they can absorb vibrations more effectively, and improve workpiece quality. Remanufacturing offers many advantages over the purchase of new machinery. In this particular example, cost savings ranged from 40 percent to 60 percent.

Sun Microsystems remanufactures and refurbishes spare parts internationally. In both Asia and Latin America, Sun brings back spares designated as FRUs, or field replaceable units. These units are repaired either in the U.S. or locally, in Latin America and Asia, depending on part value and difficulty of refurbishing the part. These parts are then restocked for reuse at a central distribution point or a RSL (Remote Stocking Location). Spares may be located in Miami, Florida for Latin America, or Singapore and other RSLs for Asia. The parts are then reused for repairs. At the time of repair, they are brought up to the current "rev" (revision) level and refurbished to new.

Hewlett-Packard also uses remanufactured parts as service parts. They are able to receive failed parts and assemblies, remanufacture and refurbish those items, and then use them as their primary materials throughout their service network.

They can reuse a valuable asset and reduce the costs associated with servicing computers and other complex machinery.

2.7 Asset Recovery

Asset recovery is the classification and disposition of returned goods, surplus, obsolete, scrap, waste and excess material products, and other assets, in a way that maximizes returns to the owner, while minimizing costs and liabilities associated with the dispositions. This definition is similar to the one that the Investment Recovery Association uses to define investment recovery. The objective of asset recovery is to recover as much of the economic (and ecological) value as reasonably possible, thereby reducing the ultimate quantities of waste.[2]

Asset recovery has become an important business activity for many companies. The importance of asset recovery to the profitability of the company depends on the ability of that company to recover as much economic value as possible from used products, while minimizing negative impacts such as environmental problems.

The attitude of many firms towards used products has been to ignore them, and avoid dealing with them after they are originally sold. Manufacturers in the United States, typically, are not responsible for products after customer use. Most products are designed to minimize materials, assembly, and distribution costs, and ignore the repair, reuse, and disposal requirements. Manufacturers have generally believed that the costs of incorporating these requirements would outweigh the benefits.

The market for products in the asset recovery process is covered in the discussion of secondary markets in Chapter 3. The research found that secondary markets both domestically and internationally are growing rapidly.

The asset recovery process can include defacing the returned products. Many retailers and manufacturers do want their brand recognizable when the products enters the secondary market. Defacing may include removing the manufacturer's name or peeling off price stickers.

Asset Recovery Supply

Firms that specialize in reclaiming value from used product enjoy a large supply of product from many different potential sources. Materials are placed into the return stream for several different reasons. One return flow type common in Europe is the result of laws requiring manufacturers to take back used products. Products coming off a lease or rental agreement make up another source of supply. Products that fail or have quality problems are another source of returned product. Quantities of this kind of return can depend on product warranties and service contracts. Forecasting return flows of defective products is often difficult. Failure or quality defect rates can depend on the type of product. For example, electronic items tend to fail early in life, whereas mechanical components fail as they age.

2.8 Negotiation

Deal-making is a key part of the reverse logistics process. In the forward flow of goods, prices are often set by brand managers and marketing specialists. Reverse logistics often includes a bargaining phase, where the value of returned material is negotiated without pricing guidelines. These negotiations may be handled loosely. In addition, one or more of the negotiation partners often does not fully understand the real value of the returned materials, creating opportunities for third parties to operate on the margin. These third parties often employ some of the sharpest logisticians. One executive told the research team that "if you want to meet a great negotiator, go talk to someone that handles scrap paper."

Sometimes the negotiations are handled by specialist third parties. These third parties act in an advisory capacity to the primary participants in the supply chain who are working to transfer ownership of the material back to the original source.

Also, third parties can manage the physical processing of the materials. Companies, such as GENCO Distribution System or DamageTrak, are examples of these kinds of firms. Generally, the same third party does not handle the value negotiations and physical processing of the product for both the retailer and the upstream manufacturer. There are exceptions to this rule, but usually retailers and manufacturers want different third parties acting on their behalf to eliminate potential conflict of interest.

2.9 Financial Management

Financial management issues are the primary determinants in the structure of a reverse logistics system, and the manner in which product is dispositioned. Most firms need to improve internal accounting processes. Accounting problems drive the actions of managers. In a few firms included in the research, accounting issues drove store managers to sidestep normal return systems. In these cases, internal policies and controls moved them to inefficient, incorrect behaviors.

An example of a policy-created problem surfaced in the research. Merchandise designated to go back to the supplier due to overstocks, or because it is not selling, is earmarked for processing through a centralized return center. However, internal accounting takes a markdown on those items that move through the centralized return center, and stores expense those items. When the centralized return center processes the material, they get full credit, and the stores are punished. The stores do not want to be punished so they slow the flow back to the vendor to postpone the negative financial impact as long as possible. This delay causes a store-level backup of material that should be dispositioned. In addition, the loss of consolidation opportunities increases transportation costs.

Often, the cost of returns is charged against the sales department. While this policy may generally be a reasonable one, it can complicate reverse logistics processes. If sales personnel are penalized for returns, they will go out of their way to slow down or demolish the quick recognition of returns and the speedy disposition of returned material. Issues related to chargebacks and bottom line responsibility for returns must be a key consideration when developing a good reverse logistics management system. As mentioned previously, the greatest roadblocks to successful reverse logistics are company policies. Generally, company policies that pose difficulties are related to accounting issues.

Sorting out what a supplier is to be paid, when deals and promotions are factored in, can be a challenge. However, returns are often the number one issue in reconciling accounts receivables. Because these issues are often so difficult to manage, third parties have begun to specialize in handling accounting and reconciliation issues.

2.10 Outsourcing Reverse Logistics

Many companies are outsourcing most or all of their logistics activities. Some of these firms are extending their outsourcing to the reverse product flow, including many firms that participated in our research. These firms are using their reverse logistics outsource supplier as a benchmark to help determine what and how reverse activities should be performed, and how much those activities should cost. Often, these outsource suppliers perform reverse activities better, and their customers find that using these service firms reduces the administrative hassle of doing it themselves. These outsource suppliers have become specialists in managing the reverse flow, and performing key value-added services, such as remanufacturing and refurbishing.

Chapter 3: Disposition and the Secondary Market

3.1 Overview of the Reverse Logistics Flow

In this chapter, product disposition options and the secondary market are examined. Once a product is initially processed, a decision is made about where to send the product: either return to vendor, to the landfill, or to the secondary market.

There are a variety of reasons why a product may enter the reverse logistics flow. They are summarized in Table 3.1 below. There are, of course, more reasons why a product will enter the reverse logistics system, but these are the most common.

Often, two identical products will follow different routes to different destinations, depending on where in the distribution channel they enter the reverse logistics flow. For example, a book that is returned to a store by a customer may not end up at the same place as a book returned by the store to its supplier due to overstocking. Neither of these books may end up in the same place as the books returned by the distributor.

When a product has been replaced by a new version, a retailer may continue to sell the old version until it is gone, perhaps at a discount. The product may never enter the secondary market. If the product does enter the reverse

Table 3.1
Reasons for Returns

Source	Reasons
Customer	1. Product did not meet customer's needs 2. Customer did not understand how to properly use the product 3. Product was defective 4. Customer abuse of liberal return policy
Retailer	1. Product packaging outdated 2. Seasonal product 3. Product replaced by new version 4. Product discontinued 5. Retailer inventory too high (overstock, marketing returns, or slow-moving) 6. Retailer going out of business

logistics flow, the firm may sell it to a liquidator for a relatively high price (by liquidator standards). This may be especially true when the new product represents only a minor, incremental improvement over an already popular product. If the product change is more significant, the manufacturer may offer the retailer more liberal incentives to sell off the remaining old product. When significant product modifications are made, the retailer may be more likely to pull the old product from its shelves, and send it to the secondary market.

When a product has been discontinued because of disappointing sales, firms are more likely to have difficulty finding a buyer for the product, even at a greatly reduced price. Retailers may attempt to dispose of the inventory, but it may be difficult through the typical retail channels. Product price in a secondary market is likely to be greatly reduced.

3.2 Returned Product Types

Retail products in a reverse logistics flow can be separated into the following categories:

1. Close-Outs: first quality products that the retailer has decided to no longer carry;
2. Buy-Outs or "lifts": where one manufacturer buys out retailers' supply of competitor's product;
3. Job-Outs: first quality seasonal, holiday merchandise;
4. Surplus: first quality overstock, overrun, marketing returns, slow-moving merchandise;
5. Defective: products discovered to be defective;
6. Non-Defective Defectives: products thought incorrectly to be defective;

7. Salvage: damaged items, and
8. Returns: products returned by customers.

Close-Outs

In a reverse logistics flow, close-out items are first-quality items that the retailer has discontinued from its product mix. In such a case, the retailer may have decided to stop carrying products sold by a certain vendor, in a particular product line.   When a firm determines that it is no longer going to carry a particular item or product line, it may contact a number of outside firms to ask for bids on removing all of the product from its stores.

Buy-Outs

Buy-outs occur where one manufacturer buys out a retailer's entire supply of a competitor's product. This purchase frees shelf space so that the manufacturer can put its product where the competitor's product was previously.   It also reduces the retailer's risk. The retailer can dispose of slow-moving merchandise, and replace it with product that will, hopefully, sell better, without incurring any cost in the transition.

Job-Outs

Job-out items have come to the end of their normal sales lives. Some products—like swimsuits and snow shovels—are only popular during a certain time of the year. When a store reaches the end of a product's sales period, the firm must either sell it at a discount, or attempt to recover some of its value through its reverse logistics system.   Unlike close-outs, a retailer is more likely to have an ongoing relationship with a job-out firm. As different products reach the end of their selling seasons, the retailer may send the products to the same job-out companies.

Surplus

Surplus is first-quality items that the company has in excess, but will continue to sell. The firm may have overestimated demand and ordered too many.

Surplus items also result from an overzealous manufacturer. This may be due to inaccurate forecasts, or because production constraints require a minimum production quantity, which is greater than the demand.

Marketing returns may also be a large source of excess product for the distributor. The distributor or the vendor may offer a special promotion, which provides the retailer an incentive to purchase a larger than usual order. If the retailer is unable to sell the product, the distributor may experience a significant increase in returns.

Defectives

Defective items have been discovered by the retailer or by the customer to be truly defective. In many cases, a firm will inform the manufacturer of the defect and the manufacturer will compensate the retailer with a new product or repayment, in the form of a check or a credit.

Non-Defective Defectives

Often, a customer will claim that a product is defective in order to return it, when, in fact, it is not defective. However, it is usually not until the product reaches a returns processing center that it is discovered to be non-defective.

Non-defective defectives often arise because a customer has purchased an item, and attempted to use it without successfully reading the instruction book. The customer then concludes that the product is defective, when in fact, it would have functioned properly had the instructions been followed.

Salvage

Salvage items have been used or damaged, and can no longer be sold as new. Salvage items lose value relative to the amount use or damage. The most difficult part of managing salvage is determining its value.

Returns

Returns are products that have been opened and used by the customer. Customer returns are generally handled similarly to salvage or surplus items. Even if a return is not defective, it usually cannot be sold as first quality.

3.3 Product Disposition

Products in the reverse logistics system are primarily disposed of through one of these seven channels:

1. Return to Vendor
2. Sell as New
3. Sell Via Outlet or Discount

4. Sell to Secondary Market
5. Donate to Charity
6. Remanufacture/Refurbish
7. Materials Reclamation/Recycling/Landfill

Depending on the condition of the item, contractual obligations with the vendor, and the demand for the product, the firm may have one or more of the above options for each item.

Return to Vendor

Retailers return products to the vendor because of defects, marketing returns, obsolescence, or overstocks. Marketing returns occur when the vendor has created an incentive for the retailer to order a larger quantity than usual, and the retailer has proven unable to sell the additional units. The retailer needs to be able to return the items.

Vendors also allow returns when they have a motivation to help the retailer avoid inventory obsolescence. In the auto industry, for example, the major auto companies allow their dealers to return a limited amount of their inventory each year. This enables the dealers to remove obsolete items from inventory, which frees space and capital to purchase additional, new inventory, and allows the dealers to better serve the customer.

Returns to vendors can also result from selling product on consignment or through a similar arrangement. In a consignment arrangement, the retailer does not assume ownership of the product. If the product does not sell, the manufacturer is usually responsible for removing the product. It is unlikely that consignment inventory will ever enter into the retailer's reverse logistics flow. A somewhat similar situation is experienced when a manufacturer sends products to the retailer with the understanding that any unsold product may be returned for full credit, except that in this situation, there is a greater likelihood that the goods may be processed through the retailer's reverse logistics system.

If a customer returns a product as defective, and the manufacturer compensates the retailer, the manufacturer may specify that the retailer must return the product. In asking for the product back the manufacturer may have two motivations. First, the manufacturer may want to identify the exact nature of the defect to determine its cause, and eliminate the defect in the future. The manufacturer may also wish to evaluate the number of non-defective defectives. By inspecting each item, the vendor gathers management information that helps to determine other disposition options. Depending on the product, the vendor may be able to reshelf and resell this product as new.

The second reason vendors want product back is to prevent the item from entering another disposition channel and cannibalizing demand. To protect the brand, the manufacturer may want to be certain that defective products are not sold again as new to unsuspecting customers. Also, to protect the brand's image, the manufacturer may not want the product to be seen in certain retail outlets, such as dollar stores or flea markets.

Another reason for wanting the product returned to the vendor is to prevent re-returns. Re-returns are product sold at a discounted price at an outlet store and then returned for full price in the regular channel.

In some cases, however, the vendor will compensate the retailer for the defective product and not require a return of the product. Depending on the vendor, the retailer may be required to destroy the product, or may be free to sell the product through an outlet store or elsewhere in the secondary market. If the retailer decides to utilize the secondary market, the vendor may require him to deface the product by removing all identifying marks and labels.

Sell As New

If the returned product is unused and unopened, the retailer may be able to return it to the retail store and resell it as new. The product may need to be repackaged. In some industries, such as automobile parts, firms spend a significant amount of money annually on repackaging, so that consumers will not be able to detect that the product is being resold.

In some industries, there are restrictions, legal or otherwise, on which products cannot be resold as new once a customer has returned them. In the building materials market, for example, in some places, it is illegal to sell a circuit breaker that has previously been installed. For this reason, if a customer returns a circuit breaker that appears to have been installed, even temporarily or unsuccessfully, the retailer cannot restock it as new.

Even when a legal violation has not occurred, companies will act quickly to put down any negative publicity that might result from an accusation of selling used products as new. For example, in 1997, an individual accused a large retailer of selling car batteries that had been previously returned. The retailer quickly responded to fight the allegation to protect the reputation of one of its best-known products.

In a highly publicized case in 1995, Packard Bell was also accused by its competitors of cannibalizing parts from used machines and putting those parts in new computers.

Sell Via Outlet or Discount

If the product has been returned, or if the retailer has too large an inventory, it can be sold via an outlet store. In the clothing industry, because customers will not accept a returned item as new, an outlet store is the retailer's only sales channel. Commonly, firms have significant quantities of end-of-season items which will no longer be offered in the retailer's stores. However, in an outlet store, customers look for, and may even come to expect, off-season items.

Selling through outlet stores offers a number of advantages over other disposition options. Using their outlet store system, firms maintain control over the products, and know where the products will be sold. For many firms, the ability to protect their reputations and market positioning is critical. However, outlet store sales also require more risk and expense.

The number of outlet stores and outlet store malls in the U.S. has tripled in number since 1988. Last year, an estimated 55 million Americans traveled at least 200 miles round-trip to shop at factory outlet stores.[1]

Many manufacturers initially opened outlet stores to sell off overstocked items, returns, or factory seconds. These stores proved to be a profitable place to sell products at a lower price point. Their success led firms to open outlet stores across the country. Van Heusen has more outlet stores than any other company with more than 700.[2]

This growth has had some unforeseen consequences. Outlet stores require inventory to remain open. Often, the overstocks from the primary retail channel and any available second quality goods are not sufficient to keep these stores well stocked throughout the year. As a result, many firms are now producing specifically for the outlet market. These items are targeted for the lower price point of outlet stores, and so offer fewer features, or are in some way differentiated from the "A" channel goods, while still offering the quality product for which the brand is known.

Outlet stores operated by manufacturers and other brand owners often provide better margins than if the product were sold to a retailer. The outlet store has become an important source of profit beyond the disposition of returns.

Sell to Secondary Market

When a firm has been unable to sell a product, cannot return it to the vendor, and is unable to sell it at an outlet store, one of its final options is to sell it via the secondary market. The secondary market consists of firms that specialize in buying close-outs, surplus, and salvage items, at prices as low as ten cents on the dollar. For retail goods, the average recovery rate is approximately 17 percent of retail price.[3] The secondary market firms then sell the product through their own stores or to other markdown retailers, such as dollar stores.

Donate to Charity

If the product is still serviceable, but perhaps with some slight cosmetic damage, retailers or vendors may decide to donate the product to charitable organizations. In this case, the retailer usually does not receive any money for the product. It may, however, be able to gain a tax advantage for the donation, and thus receive some value, while being a good corporate citizen. However, charities have begun to pay for first quality product. One large retailer included in the research charges Goodwill Industires for product.

Remanufacture / Refurbish

Before determining that the product is a complete loss, and sending it to be recycled, many firms will attempt to refurbish or remanufacture. The range of options available to a firm in this area varies greatly, depending on the type of product and the reason it entered the firm's reverse logistics flow. Many consumer products cannot be remanufactured. Once used, nothing can be done to refurbish or make them attractive or useful for another customer to purchase.

Other items lend themselves to refurbishing, such as electronics products. If a customer returns a fax machine to the retailer because it does not work, the retailer will send the machine back to the manufacturer or a third party that specializes in refurbishing. The resale value of the machine in this condition is very low. Rather than attempting to sell the machine in this condition, the manufacturer will diagnose the problem and repair the machine. At this point, the manufacturer may do one of two things with the machine: it may be sent to a secondary market firm that will sell the machine as "reconditioned" or "remanufactured;" or the machine may be sold via an outlet store.

Commonly, a defective machine is sent to the manufacturer's service network. If a customer owns a product that needs service, the product may be sent to the manufacturer for repairs. The manufacturer may offer the customer two choices: the customer can wait while the manufacturer repairs the actual product, or the manufacturer will send a different machine immediately. Under the second option, the customer does not get back the original machine.

Materials Reclamation / Recycling / Landfill

When, for some reason, the firm is prevented from selling the product to the secondary market, and the product cannot be given away, the final option is disposal. As always, the firm's objective is to receive the highest value for the item, or dispose of the item at the lowest cost. Some items, such as catalytic converters and printed circuit boards, contain small quantities of valuable materials such as gold or platinum, which can be reclaimed. Such reclamation helps offset the cost of disposing of the item. Other items may be composed of materials that are of some value to scrap dealers, like steel and iron.   When the materials are not of value to other companies, the firm may develop ways of using the product to avoid sending it to a landfill. A good example of this is outdoor running tracks made of ground-up athletic shoes. Another example involves sorting damaged retail clothing hangers, melting them, and making new hangers.

As described above, some vendors require retailers to dispose of defective products. In this case, the retailer has no choice but to follow vendor instructions and send the product to the landfill or incinerator. For example, some sports cards manufacturers accumulate inventories of cards one or more years old. Among card collectors, older cards are more valuable, and a case of cards only a year or two old may, in some instances, be worth thousands of dollars. However, the vendor may instruct the distributor to destroy the old cards, in which case the distributor, accompanied by a security guard, will place the cases of valuable cards in a landfill.

3.4 Material Flow

In any reverse logistics flow, one critical objective is to receive the highest value possible for the products in accordance with any legal restrictions or constraints imposed by the vendor. In order to accomplish this, the following steps must be a part of each reverse logistics system:

1. Gatekeeping—deciding which products to allow into the reverse logistics system
2. Collection—assembling the products for the reverse logistics system
3. Sortation—deciding what to do with each product
4. Disposition—sending the products to their desired destinations.

There are a variety of paradigms that can be used for a reverse logistics system, but most can be described in relation to the following two extremes: centralized and decentralized. In a decentralized system, all decisions regarding the disposition of products are made at retail locations. Some transportation costs are avoided because the products are not all taken to a central processing center before the disposition decisions are made. However, at the same time, a decentralized system will likely increase the total transportation costs of disposition, because all products destined for a particular secondary market firm are scattered across the company's network of stores, and either directly or indirectly, the retailer must pay to collect the goods.

The larger disadvantage of a decentralized system is that the firm will likely receive less income from the secondary market firms. There are two reasons for this. First, each location will have a much smaller quantity of a given item, and smaller quantities do not bring the highest prices on the secondary markets. Secondly, the individuals in charge of sortation at a particular store may not develop adequate experience with a particular item to learn the most effective means of disposing of the item, which means lower reverse logistics receipts for the retailer.

As described in Chapter 2, in a centralized system, all products for the reverse logistics pipeline are brought to a central facility, sorted, and sent out to their ultimate destination. This consolidation method creates the largest possible volumes for each customer, which leads to higher revenues. Employees develop expertise in certain areas and can consistently find the best destination for each product.

One of the most important activities within a reverse logistics system is sortation. Obviously, determining the best channel for disposing of the product is of critical importance in maximizing revenues from the products in the reverse logistics system. Gatekeeping is also important in reducing the system costs. Generally, a large cost is transportation. If the product is going to be thrown away, it is best to discover this as early as possible, to prevent performing additional steps and incurring additional costs. For example, if a vacuum cleaner has been damaged beyond repair, transporting it 500 miles to throw it away is not cost effective. Efficient gatekeeping early in the channel prevents such items from entering the reverse logistics system at all, and can be a source of significant cost savings.

3.5 Secondary Markets

The secondary market is a term for the collection of liquidators, wholesalers, exporters, brokers and retailers who sell product which, for one reason or another, has not sold through the primary sales channels. Companies in the secondary market sell both new and used product.

The secondary market often involves a transfer of products directly from the manufacturer to the secondary market firm; therefore, understanding the secondary market requires an examination of several areas that not part of the typical retail reverse logistics system. However, because the secondary market has a large role to play in the operations of many reverse logistics systems, a detailed discussion of these areas of the secondary market will provide a better understanding of the operations of reverse logistics flow in general.

Figure 3.1 shows a number of companies in the secondary market. To help clarify the difference between reverse logistics activities and secondary market activities, reverse logistics activities are shown in a heavy dashed line, and secondary market activities are shown in a light solid line.

If a product is entering the secondary market directly from the manufacturer, the manufacturer probably did not want the item for one of the following reasons:

1. Changed Product Packaging
2. Product Redesigned

3.　Order Cancellation
4.　Sales Expectations Not Met

Figure 3.1
Flow of Returns and Secondary Market Goods

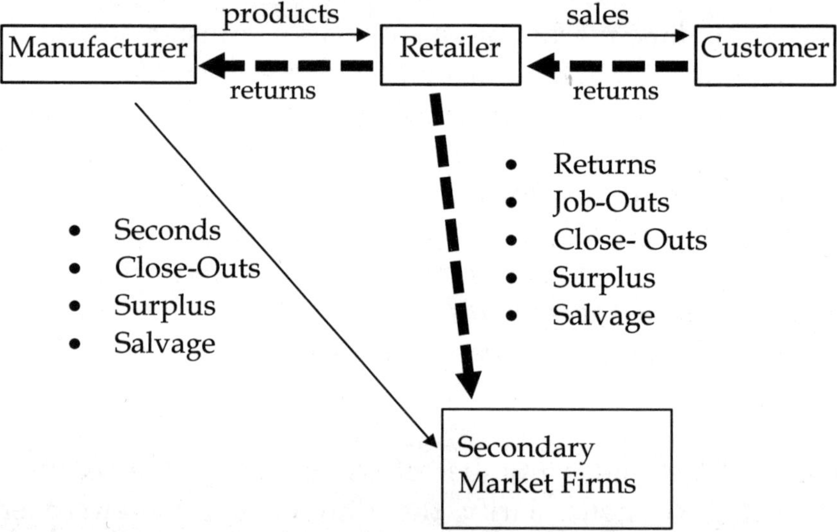

Packaging Change

Packaging changes account for a large percentage of close-out items. A package change may occur due to outdated design, or a change in the size of the product. This is particularly true with grocery items. Customers may be conditioned to a particular price for the product. Rather than increase the price of the package, manufacturers may keep the price constant by reducing the amount of product in the package, while keeping the size of the packaging unchanged.

Product Redesign

Introduction of a new product version precipitates inventory clean out of the old product. The old version of the product is sold until it is gone, perhaps at a discount. The product may never enter the secondary market. If the product change is significant, the manufacturer may give the retailer liberal incentives to sell the remaining old product. A significant modification may make the retailer more likely to pull the old product from the shelves and send it to the secondary market.

When a product has been discontinued because of disappointing sales, the firm is more likely to have a difficult time finding a buyer for the product, even at a greatly reduced price. If the product is sent to the secondary market, the retailer's expected price may be greatly reduced. Disposing of the inventory through usual retail channels may be difficult.

Order Cancellation

Cancellation of a very large order results in surplus inventory. For example, a trash can manufacturer makes a large number ordered by a major retailer. The retailer cancels the order. The manufacturer is left with thousands of surplus trash cans. The manufacturer may be quite willing to sell the unsold product to anyone willing to pay a price sufficient to recover the cost of manufacturing the goods.

Sales Expectations

Finally, a manufacturer may find itself with a large quantity of unsold merchandise because the sales projections for the product were overly optimistic. Rather than hold a large quantity of inventory that may take a long time to sell, the manufacturer may prefer to sell it all at once to a liquidator, for a lower price.

3.6 Secondary Market Firms

Firms in the secondary market supply products to a growing group of retailers who specialize in selling close-outs, surplus, seconds, and salvage items. Flea markets have long been known to sell this type of product. Now there is a growing number of discount or "dollar" stores, including national chains like MacFrugals and Just-A-Buck.

The secondary market consists of a diverse collection of companies playing different roles. These firms may be broken into several categories:

1. Close-out liquidators
    - Deal in merchandise obtained from return centers, other liquidators, directly from retail stores, or directly from manufacturers
    - Types of products: close-outs, surplus, packaging change, end-of-life
    - Physically handle inventory, sorting, consolidating, palletizing
    - Product usually remains in U.S.

2. Job-out Liquidators
    - Primarily first quality, seasonal clothing items
    - Negotiate directly with corporate buying departments of major retailers
    - Negotiate while merchandise still selling at retail store
    - End of season or transition inventory purchased a season in advance
    - Product usually remains in U.S.
3. Brokers
    - All types of products (close-out, seasonal, salvage)
    - Find liquidated merchandise for retail clients
    - No physical ownership of inventory, service provider only
    - Product very likely to be shipped outside of U.S.
4. Insurance Claim Liquidators
    - Appraisal, salvage, and recovery services for inventory losses from natural disasters
    - Primarily salvage goods
5. Barter Companies
    - Make arrangements for companies to trade a surplus of one product for another company's surplus inventory of another product
6. Gray Markets
    - New product sold by non-factory-authorized resellers
    - Products do not carry factory warranty

Because these terms may overlap considerably as they are used in the industry, these definitions may vary slightly from their common usage.

Close-Out Liquidators

Close-out liquidators deal in products that have come to the end of their shelf lives, and will no longer be sold. A product may no longer be profitable at its original price, but at a different price point, it may again become profitable. Close-outs may be created for the following reasons: the product is a seasonal item, such as snow shovels or barbecues; a new replacement product has been introduced, as with video games; the product failed to live up to sales expectations, such as toys related to movie themes.

Many close-out liquidators work directly with the manufacturer, buying the manufacturer's oversupply. A common scenario is that a major retailer cancels a large order and the manufacturer has product that it will be unable to sell to other customers. The close-out liquidator gets the product for a price which allows the manufacturer to recover its cost.

Major manufacturers prefer to deal with large close-out liquidators for two reasons: the ability to pay, and the ability to handle large volumes. A manufacturer needs to know that the liquidator will be able to pay as promised. Also, the manufacturer wants to be able to go to one liquidator for a particular deal, not to several. The large liquidators have sufficient liquidity to be able to buy in large quantities. If, for some reason, the liquidator will be unable to sell the entire product directly, the liquidator may, in turn, sell some of the product to a different liquidator. In any such sale, the liquidator has to gain a sufficient return to be able to ensure that the second liquidator cannot profitably sell the product at a lower price.

Large liquidators offer a proven track record of completing large deals that are structured to the manufacturer's wishes. In this industry, being able to deliver services as promised is extremely important. If the manufacturer wants a liquidator who can handle the entire transaction, and promises not to sell any of the product to a second liquidator, the larger firms have a distinct advantage. In the secondary market, trust is an extremely important factor in maintaining a relationship between manufacturer and liquidator.

Liquidators will often buy out the inventory of a retailer or a manufacturer at the end of the season, and store it until the following year. This is particularly true of seasonal goods, such as electric blankets, and holiday items, like Valentine's Day cards.

Close-out firms will liquidate goods in any market where they can reasonably expect to make a profit. Generally, hard goods seem to be the easiest products to profitably sell. Most liquidators included in the research merchandise hard goods. The three products that seem to be most difficult for a general liquidator to manage are clothing, toys, and electronics.

Clothing creates additional complications not found in other product lines, due to maintaining inventories of different sizes. Additionally, styles tend to change more rapidly for clothing than for hard goods. Compounded with the seasonal nature of clothing sales, this makes it difficult for liquidators to profitably buy products one year and hold them for sale the following year.

Job-Out Liquidators

Job-out liquidators are similar to close-out liquidators. They generally work with first quality items, but they deal primarily with seasonal items, not with products that have reached the end of their sales lives, or manufacturers' overstocks. Job-out liquidators often specialize in clothing and shoes. They also tend to specialize more than close-out liquidators, and try to develop product expertise and long-term relationships with retail firms.

In addition, job-out liquidators acquire goods uniquely. Typically, they negotiate directly with major retailers' corporate buying departments, while the merchandise is still selling at retail store. This technique is known as pre-selling. Finally, the job-out liquidators typically pick up the product and handle all the physical operations themselves.

Brokers

Brokers deal with all types of products (close-outs, seasonal, salvage) that have reached the end of their sales lives for any reason. Brokers are the third parties willing to pay a small price for the goods that absolutely no one else wants. It is not uncommon for brokers to enter into an agreement with a retailer to purchase anything the retailer wants to sell them for a fixed price of so many cents per pound, regardless of the type of item or its condition.

Brokers usually do not handle the product themselves. They provide the service of finding the products for outside parties who may export the products to South America, Africa or Asia, for sale on secondary markets.

In the international market, as in the domestic market, firms are generally willing to sell any product that they can turn profitably. International firms do not, for example, restrict themselves to high-value products. Generally, they are interested in anything they can profitably ship and sell.

Insurance Liquidators

Insurance liquidators specialize in products that have been declared a loss for insurance purposes. For example, if a train derails, all the product on the train may be declared totally lost, even though much of the shipment may have been completely unaffected by the derailment.

In this case, the shipper generally has two choices. The railroad company may pay the shipper in full for the loss. The railroad sells the product to a secondary market firm, like an insurance liquidator, who buys the contents of the train, and finds the best price. If the shipper does not want the products to be sold on the secondary market, the shipper can opt to receive partial compensation for the goods and get the goods back.

Barter Companies

Barter companies help firms dispose of excess products and receive in return, desired product. A barter company may have a large store of products that it has received as the result of past trades. A company gives the barter firm the inventory it wants to get rid of, in exchange for a quantity of one of the products in the barter company's inventory.

Generally, a barter company is willing to trade for any product it believes could be profitable. If the product is harder to sell, the barter company will not provide as generous a trade allowance. Some barter companies prefer to use universally valuable products—such as media advertising time or airplane tickets—as a type of currency for exchange.

A company may turn to a barter company with products whose actual value is significantly below the products' value on the company's books. Rather than writing down the value of the product, the company may prefer to trade for an equal value of media time, or some other product it needs.

Gray Markets

Gray markets sell new product outside the regular channel, usually by resellers not authorized by the factory. Product can find its way to the gray market when an authorized reseller needs to raise capital, and quietly sells still-new product to an unauthorized firm for a small profit. Gray firms typically have lower facilities costs than authorized firms, and are therefore still able to profitably sell the product at a significant discount from the manufacturer's suggested retail price.

Because gray firms are not factory authorized retailers, products sold by them do not carry the manufacturer's warranty. Gray market customers buy new, genuine product with no warranty.

3.7 Strategic Elements of the Secondary Market

The participants in the secondary market seek to achieve the following:

- Finding quality merchandise to buy and sell;
- High inventory turn rates, and
- Minimal inventory handling.

In the past, liquidating merchandise was an escape route for retailers to rid themselves of surplus and obsolete product that did not sell as anticipated. Times have changed. Effective asset recovery has truly proven to be a strategic tool for major retailers. In a competition-fierce economy where constant demand for new products is high, staying fresh means quickly selling off the old merchandise, and placing new products on the shelves. Grocery stores must operate in this manner due to perishable inventory. Highly seasonal apparel in the fashion industry makes it crucial for retailers to continually free space for the newer lines of clothing. This is necessary in order to stay competitive with others in the industry.

In the distant past, companies found it easier to dump obsolete merchandise in landfills. They saw no financial benefits from outdated merchandise. As landfill costs began to increase, companies saw the need to rethink their practices and reduce costs by sending products to secondary markets. Incorporating salvaging versus product disposal has saved one larger retailer included in the research over $6 million in landfill costs.

Efficiency in time and cost are the keys to success in the secondary market. In most cases, it is crucial for inventory to turn fast. Some firms have sufficient capital with which to buy seasonal merchandise for a large discount. They then store it for a year, and earn enough profit on the item to make up for the cost of holding the product.

Marketing the product and making it more attractive are not important once the merchandise reaches the secondary market. In most cases, consumers are aware of these products in terms of their brand name and features. Retailers in the secondary market must be aware that consumers are seeking bargains from among the products on their shelves.

Chapter 4: Reverse Logistics and the Environment

Many companies first focused on reverse logistics issues because of environmental concerns. Today, some are concerned only with reverse logistics as it relates to returning product to their suppliers. However, in the future, environmental considerations will have a greater impact on many logistics decisions.

For example:

- Landfill costs have increased steadily over recent years and are expected to continue to rise;
- Many products can no longer be landfilled because of environmental regulations;
- Economics and environmental considerations are forcing firms to use more reusable packaging, totes, and other materials;
- Environmentally motivated restrictions are forcing firms to take back their packaging materials, and
- Many producers are required by law to take back their products at the end of their useful lifetime.

Each of the trends discussed in this chapter will have significant implications for reverse logistics decision-makers.

Disposing of unwanted products is becoming a more closely monitored activity. In the United States, the traditional method of simply placing items in a landfill is neither as simple nor as inexpensive as it once was. As the number of municipal landfills in the United States continues to shrink, and as regulations affecting landfills become more stringent, the cost of placing items in landfills has risen steadily. Increased restrictions have focused on implementing or instituting greater measures to protect human health and the environment. This has resulted in the closure of many facilities, and higher costs at others. Another area of increasing regulation is the determination of what items can be placed into a landfill. Products such as cathode ray tubes (CRTs), for example, can no longer be placed in landfills.

Throughout Europe and the United States, legislation is being considered placing conditions on the legal disposition of products that have reached the end of life. In Germany, new laws dictate that the producer of a good must bear responsibility for the final disposal of the product. In some states in the U.S., similar measures are being proposed. For example, many states require retailers of vehicle batteries to take back used batteries. In places where legislation does not force a manufacturer to take back the product, products are not being allowed into landfills. In some cases, this will likely force the establishment of a system to collect these products.

Green Logistics and Reverse Logistics
An important distinction must be drawn between reverse logistics and a very related subject that we will refer to as green logistics. Reverse logistics, as we defined in Chapter 1, refers to all efforts to move goods from the point of consumption in order to recapture value. Green logistics, or

ecological logistics, refers to understanding and minimizing the ecological impact of logistics. Green logistics activities include measuring the environmental impact of particular modes of transport, ISO 14000 certification, reducing energy usage of logistics activities, and reducing usage of materials.

Some green logistics activities can be classified as reverse logistics. For example, using reusable totes and remanufacturing are both reverse and green logistics issues. However, there are many green logistics activities that are not reverse logistics related. For example, reducing energy consumption, or designing a disposable package to require less packaging are not reverse logistics activities. Designing a product to use less plastic would not be a reverse logistics activity, but designing a product to make use of reusable packaging would involve reverse logistics.

In this chapter, several issues that lie at the intersection of green logistics and reverse logistics will be discussed.

4.1 Landfill Costs and Availability

Landfill Technology

Waste disposal has not changed dramatically since the fifth century B.C., in ancient Greece, where people were responsible for carrying their own garbage to the town dump. During the time of the Roman Empire, people in cities shoveled their garbage into the streets, where it was collected by horse-drawn wagons, and taken to a centrally located open pit. Dead animals and people were placed in a pit outside of the city due to pungent odors. These methods lasted only as long as the Roman Empire. They collapsed during the Dark Ages, and were not yet reinstituted during the Renaissance. During these times, trash was generally discarded without much thought given to its effect on people and the environment. A survey done in 1880 showed that only 43 percent of major American cities had some minimal type of garbage collection. By the 1930s, this number had risen to 100 percent.[1]

Until the 1950s, waste disposal still consisted primarily of burying waste in a large pit. Spontaneous combustion sometimes occurred. Sometimes the waste would intentionally be burned to reduce volume. By the mid-1950s, the need to analyze groundwater downgradient from landfills was recognized, and by 1959, the sanitary landfill was the primary waste disposal system used in American communities.[2]

In a sanitary landfill, also known as a controlled tip, refuse is sealed in cells formed from earth or other materials. The modern landfill is different from the dump of the past in the way the material is covered. By the 1950s, it was recognized that covering the material with a layer of soil would reduce its attractiveness to animals, and reduce odor problems. Often, this layer was not deep, and plants were encouraged to grow on it. As studies continued of landfills' effects on water quality, it was learned that rain and other water were seeping through the landfill, contaminating the water table. This contaminated water is called leachate. To reduce water contamination, caps of a nonporous material are now placed on top of the landfill to reduce the amount of leachate generated. Also, a liner of nonporous material is added to the bottom of the landfill to reduce the amount of leachate escaping. In the landfill of the 1990s, systems are used to collect leachate at the bottom of the landfill, before it slowly seeps out through the nonporous material. To capture methane and other gases produced by decomposition, collection systems must be put into place.[3]

Landfill Availability

For a number of years, there has been a perception of an impending shortage of landfill space.[4] In 1988, it was believed that nearly half of the metropolitan cities on the East Coast would have no further landfill capacity by 1990.[5] Although this claim has not been borne out, there has been a steady decline in the number of landfills.

In 1986, the Environmental Protection Agency's Office of Solid Waste developed a list of municipal solid waste landfills (MSWLFs). Municipal solid waste includes household waste, and some industrial wastes, such as office paper and pallets; but does not include construction waste, car bodies, and industrial process wastes that might be disposed of in landfills.[6] This was the first time such a list had been compiled. In this original list, 7,683 MSWLFs were listed. When the list was updated in 1992, the number of landfills had declined to 5,345. When the list was revised in 1995, the number of municipal solid waste landfills had declined to 3,581. However, it should be noted that during this period, the new requirements for proper cover for a closed landfill went into effect, and this regulatory change led to the closure of many landfills.[7]

Although the number of landfills has been shrinking at a significant rate, it is the actual amount of landfill space available, and how long it will last that is the primary concern. Many smaller landfills closed because they were not able to afford the cost of being compliant with new regulations.[8] According to the EPA's Municipal Solid Waste Factbook, 29 states have 10 years or more of landfill capacity remaining, 15 states have between five and 10 years of landfill capacity remaining, and six states have less than five years of landfill capacity remaining.[9]

Obtaining information on the rate at which new landfills are being created is difficult, but as to whether there is a landfill crisis, the EPA has a short answer. Despite the fact that the number of landfills has been decreasing, the capacity has remained "relatively constant.... New landfills are now much larger than in the past."[10]

Cost of Landfill Usage

Along with the perception of rapidly depleting landfill space, there has also been a perception of rapidly rising prices for landfill usage. In the industry, the standard basis for comparing the cost of landfill is the tipping fee. A tipping fee is the cost charged to dispose of a ton of waste.[11]

As we have seen above, the perception of rapidly depleting landfill space has not been borne out by the facts. However, in the case of landfilling costs, the perceived trend is very real. According to the EPA, the national average tipping fee in the United States increased from $8 to $31.50 from 1985 to 1996, an increase of 294 percent. This is an annual growth rate of 9.4 percent.

Since 1985, the Pennsylvania Department of Environmental Protection has collected data on tipping fees in Pennsylvania, which show an almost identical trend. According to their data, the average tipping fee in Pennsylvania has grown by 300 percent between 1985 and 1996, increasing from $11 to $44 per ton.

Although nationwide prices and some regional prices show a long, slow, upward trend, in some regions, prices have gyrated wildly. After declaring bankruptcy in 1994, Orange County tried to raise revenues by increasing tipping fees from $23 to $35. Because there is excess dump capacity in the region, many of its customers took their garbage to cheaper dumps elsewhere. To increase business, the rate was subsequently dropped to $19 per ton.[12]

The fluctuating cost of landfilling has also caused plans for huge new household waste dumps to be dropped after tipping fees failed to stay above the projects' breakeven points.[13]

Garbage Generation
Even though there is no immediate threat that garbage will begin to pile up in the streets, landfills and landfill usage are going to be important issues in the near future. Americans

generate garbage at an astounding rate. Solutions must be developed.

A few facts about Americans' ability to generate garbage:

- Every year Americans use 75 billion disposable paper cups.[14]

- King Khufu's great pyramid in Egypt is built of more than 3 million cubic yards of stone. The Great Wall of China is built of 120 times as much material.[15] The Fresh Kills Landfill on Staten Island, New York, is run by the New York City Department of Sanitation. According to estimates, in the very near future, the total volume of the landfill will exceed that of the Great Wall of China.[16]

- The current rate of garbage production in the United States is hard to comprehend. The average American generates 4.34 pounds of garbage per day.[17] According to a 1987 study, the then-current U.S. annual waste generation was 228 million tons, which is an amount sufficient to cover an area 654 miles square 10 feet deep.[18]

- The EPA projects that from 1995 to 2000, the annual rate of garbage generation per person in the U.S. will increase from 4.34 to 4.42 pounds per person per day; a modest increase of two percent. Because of increased recycling efforts, however, the amount heading for landfills is expected to decrease over this period, from 2.47 pounds per person per day to 2.38 pounds, a reduction of 3.6 percent.[19]

Given the rate at which Americans generate waste, landfill alternatives must be developed.

Extending Landfill Capacity

With increasing regulations on the location, construction, and operation of landfills, opening a new landfill has become difficult. The "not in my backyard" (NIMBY) sentiment among people living in the location of a proposed new landfill adds to the dilemma. Additionally, the NIMTOO, "not in my term of office," phenomenon makes the problem worse.[20]

As creating new landfills becomes more difficult and more expensive, communities are trying to find new ways to prolong the lives of their existing landfills by reducing the volume of material that goes into them.

It appears that the amount of garbage being sent to landfills is shrinking at a rate faster than the rate of population growth. According to the EPA, the number of tons of garbage sent to landfills (in millions of tons) in 1970 was 87.9. The number grew to 132 million tons in 1990, but had declined to 118 million tons by 1995.[21] There are also some regional indications that the amount of garbage being sent to landfills is decreasing. In 1990, California's landfills disposed of 40 million tons of garbage. In 1996, that number had been reduced to 32.8 million tons.[22]

In 1970, 73 percent of all waste was sent to a landfill. By 1995, that number was down to 57 percent.[23] The reduction in material sent to the landfill has been achieved by increasing the amount of material that is dealt with in other ways, through recycling, composting and incineration. Although the amount of material that is incinerated has grown steadily over this period, as a percentage of the waste stream, it has declined from 20 percent to 16 percent.

The decline in landfill and incineration is a result of large increases in the percentage of the waste stream that is being recycled or composted. In 1970, less than seven percent of the waste stream was recycled or composted. In 1995, 27 percent of the solid waste stream was recycled or composted.[24] Of this, roughly five percent of all waste in the United States is composted, and roughly 22 percent is recycled. Nearly all states have ambitious goals of the percentage of waste that should be recycled. Rhode Island has the highest long-term recycling goal, at 70 percent.[25] Recycling programs are available in the major cities in all states.

Landfill Restrictions

Restrictions on what can be placed in a landfill are a key factor to the longevity of America's landfills. Some items, like CRTs, are banned because placing them in a landfill will present a long-term health risk. Others are banned because they take up too much space and can be better dealt with using other methods.

As of 1993, 44 states had bans on telephone books being placed in landfills, and 23 states had laws banning grass clippings and yard waste from landfills.[26] These are but two of the many items which have been forbidden from being placed in landfills because they can be dealt with so much more effectively in other ways. Telephone books can be recycled. Yard waste can be composted. However, either of these options becomes impossible once the products have entered the landfill. Agencies charged with collecting waste have set up additional systems to collect materials that have efficient alternate waste disposal possibilities.

Disposal Bans and Reverse Logistics

In many places, the definition of a hazardous material is being expanded. In Minnesota, for example, it has been ruled that automotive shocks and struts cannot be placed in landfills, and it is expected that other states will follow suit in the near future.[27] For automotive shocks and struts, this new restriction has resulted in a perceived new opportunity. One major manufacturer of automotive shocks and struts has begun a program to collect used shocks and struts and reclaim the materials contained in them. Motivated by the landfill ban, the manufacturer began studying the feasibility of collecting the used shocks and struts. Once the study was begun, the company quickly concluded that a substantial market existed for some of the specialized steel used in the manufacture of the shocks and struts. The company also realized that such a program would give it significant, beneficial exposure as an environmentally friendly company. On this basis, the company is proceeding with the implementation of a nationwide program to reclaim the products.

Various computer components are banned from landfills, including circuit boards with high lead content. The computer components representing the largest problem are the cathode ray tubes (CRTs) in computer monitors. CRTs, which have been banned from landfills by the EPA since 1992,[28] contain traces of lead, phosphorous, cadmium, and mercury. When CRTs are crushed at trash-compacting facilities, the lead and other particles become airborne, posing a health risk to sanitation personnel and those living nearby.

Experts believe that the magnitude of this problem will only increase. The Gartner Group predicted that between 1992 and 1996, 50 to 70 million personal computers would be discarded. A Carnegie Mellon study predicts that 150 million personal computers will be heading for landfills by 2005, because they cannot easily be recycled.[29] The disposal cost for these machines alone will exceed $1 billion, excluding the cost of creating a hole large enough to contain them. If a one-acre hole were dug to bury this many personal computers, it would be three and a half miles deep; enough room to stack about 15 Empire State Buildings end to end.[30]

The size of the future PC disposal problem can also be considered in the following way. Currently, there are 324 million PCs in use around the world. If all of these were placed in a landfill one acre square, the hole would need to be 6.7 miles deep, nearly as deep as the Mariana Trench.[31]

Pallets are another landfill-banned product. It has been estimated that there are 1.6 billion pallets in the United States, enough for every American to have six pallets. This year alone, nearly 400 million pallets will be produced, which requires a tremendous amount of wood. It also poses a serious disposal problem. A third of U.S. landfills will not accept pallets, and others charge extra fees for disposing of pallets. Burning the pallets is not necessarily the answer, either. In Wisconsin, a large building products company was fined $1.7 million for illegally disposing of incinerated pallet waste.[32]

These are but a few of the products banned from landfills. Across the country, the following products have been banned from landfills in one or more states: motor oil, household batteries, household appliances, white goods, paper products, tires, thermostats, thermometers, fluorescent lights, and some medical and electrical equipment.[33] The number of products prohibited in landfills is only expected to increase in the future. As in the case of automotive struts, each one of these bans will present a new reverse logistics opportunity.

4.2 Transport Packaging

Although it does not receive much attention in the trade press, the usage and impact of transportation packaging, pallets, drums, gaylords, boxes, etc., is a significant portion of the total packaging used globally each year. As described in section 4.1, many governmental bodies are placing restrictions on the ways that transport waste may be disposed. As is described in Chapter 5, companies in

Germany are responsible for taking back all of the transport packaging of the products they sell.

In the United States, legislation does not exist nationally, and very few statutes concerning transport packaging exist. However, transport packaging is a significant issue for many companies. Consider a few facts:

- Of the 1.6 billion pallets in the United States, each year, pallet recyclers will process more than 170 million pallets for recycling, representing 2.6 billion board feet, of which 2.3 billion board feet will be reused to make new pallets.[34]

- Although wood is the most common material for pallet construction, plastic pallets currently represent about 3 percent of the market, but are estimated to be growing by as much as 30-40 percent per year.[35]

- Each year, $15 billion worth of corrugated fiberboard is sold in the U.S. alone, generating more than 24 million tons of waste.[36]

The difficulties of pallet disposal were discussed above in the section 4.1. In some U.S. cities, the cost to dispose of corrugated can be as high as the cost of purchasing it.[37] The rising cost of purchasing and disposing of transport materials is leading many companies to consider reusable transport packaging.

Reusable Transport Packaging

Although environmental reasons have factored heavily in many firms' switch to reusable transport packaging, for some, it is simply a matter of economics. Reusable containers are generally much more expensive than single-use packaging. However, if a reusable container is reused a number of times, the per-trip cost of the reusable container quickly becomes less expensive than the disposable packaging.

A number of case studies cite companies' significant savings from using environmentally friendly packaging. For Johnson & Johnson, the payback period on reusable gaylords, both domestic and international, for inter-plant shipments was three to six months. John Deere & Co. experienced a two-year payback on its reusable crate program, which has subsequently been expanded to its retail outlets in the U.S.[38]

Although many case studies have discussed companies that successfully implemented reusable container programs, few cases have been completed about companies that investigated the use of reusable containers and determined them to be uneconomical. One interesting example is the Harley-Davidson Company's decision not to use returnable crates for delivering motorcycles to dealers, after exploring the possibility of their use.[39]

Reusable Container Types

Reusable shipping containers are available in many shapes and sizes and many different materials. Custom containers can be designed to suit any particular need.

Over the years, corrugated boxes have set the standard for all other packaging. Corrugated is lightweight, strong, easy to handle, and inexpensive. Therefore, it is not surprising that the majority of reusable shipping containers is plastic, wood, or metal replacements for corrugated boxes.

Plastic

In many applications, plastic is the lightest and cheapest option. Plastic reusable containers generally come in two styles: rigid and collapsible. Rigid plastic containers come in a wide variety of sizes, but are most commonly found in sizes slightly larger than a box of copy paper.

Rigid plastic containers come in two primary types: with or without integral lids. Those without integral lids can be stacked in a nested fashion, as shown in Figure 4.1. This reduces storage space and transportation costs. Integral lids often consist of two plastic flaps, one on either side of the opening, that interlock in the middle when the box is closed, as shown in Figure 4.2. This provides added protection for the contents. When the box is open, the flaps hang down on the outside of either side of the box. Unfortunately, these flaps can be obstructive and cause the containers to turn in the material handling system, if the system is not properly designed to handle the containers.

Figure 4.1
Rigid Totes

(photo courtesy of CHEP USA)

Figure 4.2
Rigid Totes with Integral Lids

(photo courtesy of Buckhorn Inc.)

Collapsible plastic containers, shown in Figure 4.3, are also available in a wide variety of sizes, but the most commonly used is the size of a pallet. When standing, these containers are the size of a typical gaylord. The four sides fold down flat, which leaves the container the size of a very thick pallet, as shown in Figure 4.4. The sides are typically held in place in the upright position with snapping locks. The bottom is of pallet-type construction to allow easy handling with a forklift truck.

Figure 4.3
Folding Plastic Bulk Containers

(photo courtesy of ROPAK Corp.)

Wood

Wooden containers are most commonly found in pallet-sized gaylords, as shown in Figure 4.5.. As with large plastic

Figure 4.4
Folded Plastic Bulk Container

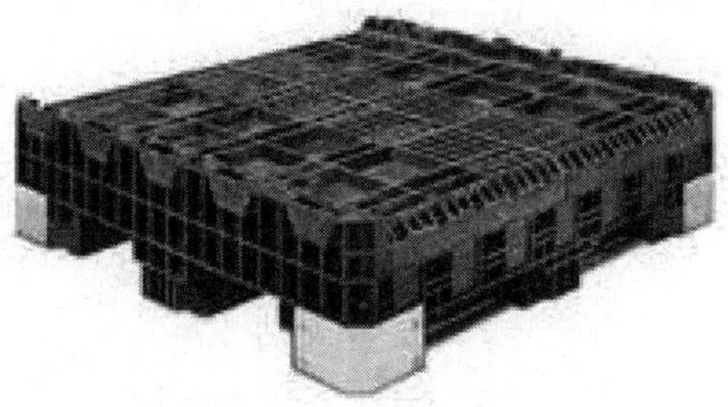

(photo courtesy of Buckhorn Inc.)

Figure 4.5
Collapsible Wooden Bulk Container

(photo of NEFAB RePak T courtesy of NEFAB, Inc.)

containers, the bottom is of pallet construction. Instead of folding the sides down to collapse the container, the sides are removed as a unit from the top and bottom. The sides are hinged together at the corners, so they can fold flat, as shown in Figure 4.6.

Figure 4.6
Disassembled Crates Fold Flat

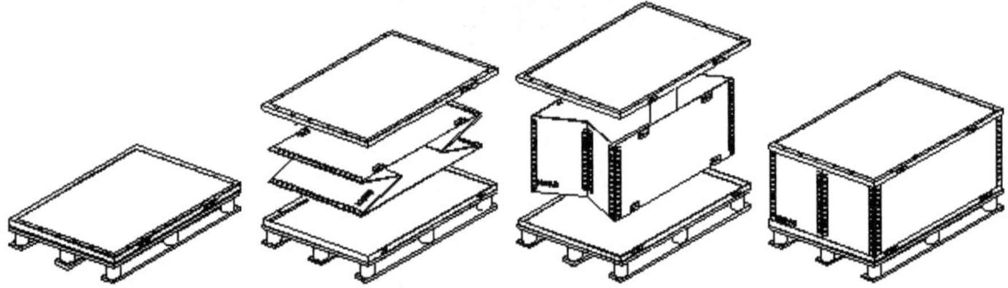

(illustration of NEFAB RePak L courtesy of NEFAB, Inc.)

Pallet collar containers are assembled by stacking layers of collars onto a pallet base, as shown in Figure 4.7. Using different numbers of sections, the container can be made as short or as tall as needed. These containers also have an ergonomic advantage, in allowing the container to be disassembled as the product is removed.

Wooden containers are very durable and strong, and are suitable for heavy loads and rough handling. Unfortunately, this sturdiness comes at the price of being heavy, which adds to transportation costs and can make them more difficult to maneuver. However, for products that require

Figure 4.7
Pallet Collar Containers

(photo of NEFAB RePak P courtesy of NEFAB, Inc.)

sturdy transportation packaging, (like machinery, scientific equipment, etc.) this increased reverse transportation cost is more than made up for in avoided damages and long-term packaging cost savings.

Metal

Finally, metal is often used for constructing wheeled cages that are used to carry boxes or totes for delivery. These cages are approximately six feet high and three to four feet long, by two feet deep; large enough to carry many totes and other miscellaneous items. If all items for a given customer are placed in one or more cages, loading and delivery times can be significantly reduced. However, because they do not collapse or nest for return transport, they are primarily used for local deliveries, not long-distance transportation.

Metal is also used for gaylord-sized bulk containers. In size and shape, these are similar to the plastic bulk containers shown in Figure 4.3.   Unlike the plastic containers, they cannot be collapsed, and given their size and weight, they generally cannot be nested. Given their steel construction, however, they are more durable than plastic containers. Like wooden containers, these bins are strong, durable, and heavy.

Because of their weight and poor transportability, metal bulk containers are better suited for storage than for transportation. They are typically used for storing heavy, loose parts.

Container Pools

Given the high initial investment required for reusable containers, being able to get the containers back is critical. If a reusable container program is designed to operate only within the area near a facility, getting the containers back is not difficult. Transportation costs should not outweigh the material savings from the program.

However, if the partners are far-flung, the reverse transportation costs may be prohibitively high, and render the program uneconomical. Ensuring that the containers are returned may prove more difficult as the distance increases. Although an in-house closed loop system is easy to maintain when transporting locally, it is much more difficult to make such a system work on a national or international level.

To maintain the supply of containers, firms often charge a deposit on them when they are initially sent.[40]   For this reason, there are arrangements that allow companies access to a large pool of containers. The company does not purchase the containers, but pays a fee for their use. The fee typically is based on some measure of the firm's use of the system. For example, a fee is charged every day for every container the company uses.

The best known such program is run by CHEP. CHEP owns millions of pallets, each painted bright blue. Users of CHEP pallets pay a fee of 3.5 cents for every day that they have a CHEP pallet in their facility. This may prove exorbitant for use in long-term storage, but for long-distance transportation, it can represent significant savings. Once the pallets are delivered to another company, the sender is no longer responsible for the cost of the pallets. The shipping company gets all the use it needs from the pallet for less than a dollar. This is much cheaper than paying as much as $10 to purchase a pallet.

Pallets are not the only transportation materials to be used in large pools of this type. United States railroads have a similar arrangement with boxcars. A stock of cars is pooled, and the different railroads use them as needed. The pool is jointly owned and managed.

Dunnage

Once a decision has been made about the construction of the container, a decision still must be made about the dunnage to be used. Dunnage is the material used inside the container to prevent damage during shipment. Dunnage may be plastic, polystyrene, paper, corrugated, or other materials.

Polystyrene

Many products (especially electronics) are packaged in polystyrene foam (styrofoam) to prevent in-transit damage. Because of its inorganic nature, polystyrene has long been a target of environmentalists. In the early 1990s, several states and 40 local governments enacted bans or recycling requirements on the use of polystyrene. However, over time, many of these restrictions have been eased.[41]

Although environmentalists may dislike polystyrene foam, they do not have science on their side. One study found that polystyrene "peanuts" consume less energy to manufacture, produce less air and water emissions, and produce less solid waste than other packaging alternatives, including paper.[42]

Recycling rates for polystyrene foam nationwide have remained around 10 percent, and have not reached anticipated levels. The inability to find a workable method for recycling the polystyrene peanuts is expected to play a large role in determining the future of polystyrene foam. It is expected that polystyrene foam demand will experience slowing growth as overall demand for disposable packaging decreases.[43]

Despite the fact that restrictions may be easing in the United States, worldwide this may not be the case. In June 1997, Beijing, China mandated that manufacturers and sellers of polystyrene food boxes must recycle one-third of their production by 1998 and two-thirds by 2000. As a result of producer responsibility mandates either in place or under consideration in 28 countries, electronics firms have been turning to paper cushioning, because of high collection fees on plastics.[44]

A number of companies have attempted to overcome the drawbacks associated with polystyrene foam. Starch-based peanuts are one alternative, but they often turn into "mashed potatoes" in humidity.[45]   Popcorn, although biodegradable, may leave an oily residue. A company in California has introduced a product made from limestone, recycled potato starch and water, which it claims biodegrades, but does not face the problems experienced by the other products. The McDonalds restaurant chain has approved the product for use with its Big Mac sandwiches and has agreed to purchase 1.8 billion of the containers beginning in late 1998.[46]

4.3 Returnable Packaging Considerations

Benefits of Reusables

One of the major reasons why companies consider reusable transport packaging is to save on the purchase and disposal costs of disposable packaging. As described above, many local governments have instituted prohibitions on disposing of corrugated material in landfills. This means that companies must store the corrugated material somewhere, and arrange for an outside company to collect and recycle the material.

In addition to a lower per-trip cost, reusables may provide better protection for the products being shipped. Reusables can provide the user with much future flexibility. As transportation requirements change, reusables can often be reconfigured by modifying the dunnage, which is much cheaper than buying new containers. Finally, if the company no longer has any use for the containers, they can be returned to the vendor for a credit. Because containers are often made of recyclable materials, they are recycled when they have reached the end of their useful lives.

Many companies enter into a program of using reusable packaging because they believe it is environmentally correct. While this may be a noble motivation, it may not always lead to sound business decisions. As will be shown below, there are many costs and issues that must be considered before switching to returnables.

Transportation Costs of Reusables

Transportation materials costs are by no means the only consideration in a decision about reusable containers. Many of the company's costs related to handling, transporting, and tracking shipments and materials will be heavily affected by a change to reusable containers.

Transportation costs are a major stumbling block to reusable containers. They tend to be heavier than the corrugated materials they replace. Because shipping costs can be weight-related, this translates into higher outbound transportation costs. If trucks "weigh out," that is, they are filled to their maximum weight limit, the extra weight of reusable containers means that fewer units can be put on each truck, which also means higher shipping costs.

However, in some cases, reusables reduce transportation costs. When one major manufacturer switched to returnable containers for appliances, the containers were strong enough to be double-stacked, unlike the expendable corrugated cartons they replaced.

In addition to adding weight, "cube" utilization may not be as good with reusables. A company will typically only invest in a relatively small number of different sized containers. Disposable materials may offer a wider variety of sizes. The result is that reusable containers may contain more wasted space. This translates into more truck space to move the same amount of product and higher costs. Also, standardized sizes can mean that more dunnage material will be needed, again raising costs.

The other major transportation cost of reusables is getting the containers back to the company. If the containers are taken to the customers on company trucks running dedicated lines, bringing the containers back should have a negligible impact on transportation costs. When this is not the case, the cost of bringing the containers back may be high, and in some cases, high enough to make the use of reusables uneconomical.

Ergonomics

Ergonomics must play a role in container selection. Some materials are much heavier than others. A gaylord-sized wooden container can weigh over 100 pounds, which is too heavy for a single person to comfortably lift on a regular basis, which means two people or a forklift are needed. Also, the contents of some larger sizes cannot comfortably be reached without repositioning the container.

Reusable containers often offer an ergonomic advantage over disposable containers. Most collapsible plastic gaylords offer a drop-down panel in one side of the container, as shown in Figure 4.3. Once the level of parts in the container drops below the bottom of this door, the container can be opened. This makes it easier for employees to reach parts in the bottom of the gaylord, reducing physical strain.

Other Costs of Utilizing Reusables

Although many companies consider materials and transportation costs, many fail to adequately consider all of the other costs involved in a returnable program. In addition to the costs of sending and getting back the containers, other handling costs include cleaning, repairing, storing and sorting the containers. Costs to consider are summarized in Table 4.1.

Table 4.1
Costs of Utilizing Reusable Containers

- Forward Transportation Costs
- Reverse Transportation Costs
- Container Inventory Management
- Inspection
- Cleaning
- Repair
- Storage
- Sorting
- Adapting for Future Use

Unfortunately, reusable containers do require some maintenance. Once the containers return, they may need to be cleaned before they can be used again. Some types may need to be inspected before they are reused, to prevent the use of damaged containers. When damaged containers are discovered, either at an inspection station or on the line, provisions for repairing or replacing the damaged containers are needed. As containers reach the end of the useful lives, replacements will have to be purchased.

For both the producer and the consumer of the containers, the cost of storing the containers must be considered. As the source of the containers, the company implementing the use of reusables can expect to maintain the supply of empty containers.   Somewhere, space must be found for this purpose, which would lead to an increase in costs. Also, additional labor will be incurred in storing and moving the containers. The consumer must have space to store empty containers waiting to be sent back to the producer.

Managing Containers

One unpredictable cost is tracking the containers. In theory, the task should be easy: containers go to the customers, and they return. If there is only one customer that the containers are sent to, this task may be easy. However, if that customer is also receiving reusable containers from other supply chain partners, there is a very real chance that the containers may be kept or stolen. A dairy in Southern California retains a private investigator to find and capture milk crates.

Even when the customers are part of the same corporation, the customers may not have any incentive or motivation to help the supplier get the containers back. Many large retailers ship products to their retail stores from distribution centers using the most common reusable shipping material, the lowly pallet. The retail stores are supposed to save the pallets they receive and send them back to the distribution centers.   However, the distribution center managers interviewed were unanimous in agreeing that it is rare that a store collects and returns the pallets. Instead, most are landfilled, given away, or sold.

To keep track of containers, many companies use bar coding to track individual containers.   Other companies are developing specialized information systems strictly for this purpose. For example, GENCO Logistics has developed a stand-alone software package that traces individual reusable containers for Wal-Mart. Wal-Mart uses the system to track containers that are used to transfer returned goods from stores to a returns processing center. Using the system, Wal-Mart is able to track the progress of an individual container from the store, through the carrier, until it reaches the returns center.[47]

As this example illustrates, companies are beginning to realize that there can be benefits to enlisting a third party company in helping to manage reusable containers. Some companies, like Wal-Mart, require a software package to help them manage their stock of containers. Other companies are turning their container management over to third party companies, entirely. The idea of allowing a third party to manage a company's stock of reusable shipping materials is not new. For many years, CHEP has been maintaining a stock of pallets used by companies around the world.

Although many reusable container projects prove to be beneficial, this is not always the case. In 1995, a study at the Amsterdam Free University traced the use of reusable plastic totes in the dry grocery goods distribution industry. Currently, plastic totes are the standard means for products shipped from manufacturers to distributors, and from distributors to retailers. The study looked at all system costs, and concluded that using one-way cardboard containers is the most efficient way to distribute products.

The authors explain that the cost of maintaining the necessary pool of containers, storing, cleaning, checking for damage, etc., are large enough to offset any long-term lower cost of the units.[48]

Success Factors

Although no two reusable container programs are alike, there are a number of factors that have a significant impact on the likelihood of success of a given program.

1. Transportation Distances. The shorter the distance that containers are hauled, the lower the cost of the program. Shorter distances will obviously reduce transportation costs. Also, shorter distances mean fewer containers in transit at any given time, which translates to reduced need.

2. Frequent Deliveries. The shorter the time between deliveries, the fewer containers will then accumulate between trips. The fewer containers piling up at either end of the relationship, the fewer containers that need to be purchased, and the less space that will be needed for storage. Also, the longer those containers spend gathering dust at the customer, the greater the opportunity for damage and losses.

3. Number of Parties Involved. The fewer parties involved, the easier it is to keep track of containers, and the less opportunity for lost containers. To manage a system with many partners, some companies try to assign each container to a particular partner, and maintain a separate

closed-loop with each partner. This makes it easy to know which partner has a particular container, but creates a number of other administrative problems.

4. Number of Sizes Needed. As the number of different sizes of container increases, better cube utilization can be obtained, which leads to lower transportation costs. Unfortunately, using many container sizes generally means that more containers must be purchased, to guarantee availability of the needed size. In addition, more containers must be handled and stored at each location.

5. Partner Buy-In. The other half of the relationship may incur a significant amount of additional work as a result of a change to reusable materials. If the consumer is also sending material to the supplier, the consumer may benefit significantly from the change.

4.4 Product Take-Back

A number of societal changes regarding the environment are having a profound impact on reverse logistics, as shown above. In general, there appears to be a trend toward greater restrictions and limitations on what may be placed in a landfill, as well as how and where a product may be disposed.

As a result of these changes, companies have begun to examine new ways to regain value from products once they have reached the end of their useful lives. In some cases, the impetus is strictly economic, as with companies that reclaim the copper and other valuable materials from electronic components.

In other cases, a change in environmental laws can alter the economics of recovery. In the landfill section, a strut manufacturer was discussed. Legislation was passed in one state preventing their product from being landfilled. This led the manufacturer to profitably recover the materials in its product.

Finally, in some places, laws force the manufacturer to take responsibility for proper disposal. In some cases, legislation mandates that manufacturers must be willing to take back products from consumers after the products have reached the end of their lives.

Take-Back Laws in the United States
Some companies have begun to realize the potential marketing benefits of a take-back program. In the U.S., the President's Council on Sustainable Development has begun to study the idea of Extended Product Responsibility (EPR). Extended Product Responsibility focuses on the total life of the product, looking for ways to prevent pollution and reduce resource and energy usage through the product's life cycle.[49]

The President's Council on Sustainable Development endorsed the general principle of EPR and said that current voluntary programs seem to be working well.　It recommended the adoption of a voluntary system of EPR.

Many companies, such as Compaq, Hewlett-Packard, Nortel, Frigidaire, and Xerox have adopted EPR.　Also, industry-wide programs have been created to recycle nickel-cadmium batteries and auto bumpers.[50]

Take-back programs in the United States are not prevalent. However, some are developing.　At least 15 states have laws requiring retailers to take back vehicle batteries.　Maryland passed a law requiring manufacturers and retailers to set up a system for collecting mercury oxide batteries.[51]

Advanced disposal fees (ADFs) are paid by the consumer at the time of purchase to cover the cost of disposing of the product at the end of its life.　At least 22 states have ADFs on tires, many states have them on motor oil and some have them on white goods, such as appliances.

Computers

While there is currently no mandatory take-back of computers in the U.S., the U.S. is the world's leading producer and user of personal computers.　The U.S. electronics industry has begun to study how to recycle electronic products,[52] and has begun designing an ideal electronics recycling center.[53]　The U.S. EPA has begun studying the collection of end-of-life electronics components. So far, it has funded two collection pilots for residential electronic and electrical equipment.[54]

In Japan, by the year 2000, makers of electrical devices will be required to recycle their own products. In response to this coming deadline, IBM Japan began a program to encourage customers to trade their old computers in on a new system. Depending on the age and value of the system being brought in, the customer receives a certain amount of credit. The returned computers will have their processors and hard disks upgraded, and will be returned to stores to be sold.[55]

It is estimated that in California alone, more than two million obsolete PCs are abandoned every year. Many charitable organizations have created standards for acceptable and unacceptable computers. In the words of a spokeswoman for Goodwill Industries, a large U.S. charitable organization, "We don't want your junk."[56]

Chapter 5: European Reverse Logistics

As in the U.S., effective management of reverse logistics is still emerging in Europe. In environmental and green issues, Europe appears to be ahead of the United States. For consumer returns, European reverse logistics practice appears to lag behind leading edge American systems.

The focus of this chapter is on European environmental and green solutions to discarded products and materials.

Throughout Europe, legislation is being passed, placing conditions on what can and cannot be done with a product that has reached the end of its life. For example, a number of European countries have passed legislation requiring producers to collect their products at the end of life. Many believe that it is only a matter of time before similar measures appear in the U.S.

5.1 German Packaging Laws

The German Packaging Ordinance of 1991

No single piece of packaging legislation has been as widely discussed, nor has had as wide an impact as the packaging laws recently implemented in Germany. The German system is worth examining at in detail because it has been so widely discussed in the popular business press, and has formed the basis for programs in other countries. It is the motivation for pending legislation in the U.S.

In 1991, the German *Bundesrat* (the German equivalent of the U.S. Senate) passed the Packaging Ordinance. This legislation mandates that industries organize the reclamation of reusable packaging waste, while local authorities continue to handle the collection and disposal of the remaining waste. Under the legislation, companies must either collect the packaging themselves, or contract with someone else to perform the collection.[1]

According to Ackerman (1997), the 1991 Packaging Ordinance has four principal components designed to increase producer responsibility for packaging waste:

1. Manufacturers and distributors must accept transport packaging, such as pallets, cartons, etc., and reuse or recycle them.

2. Retailers must accept back secondary packaging, for example, outer packaging, like the box that a tube of toothpaste comes in. Distributors must accept back secondary packaging and reuse or recycle it.

3. For primary packaging, such as a toothpaste tube, the same rules apply as for secondary packaging, unless the industry establishes a collection and recycling system that meets strict governmental quotas for the recovery of each type of packaging (72 percent for glass and metal packaging, 64 percent for paper, plastic, cardboard, and composite packaging).

4. A deposit/refund system is required for beverage, detergent, and paint containers.[2]

The legislation quickly had its desired effect. For transport packaging (which is 30 percent of the packaging), interest in reusable packaging quickly increased, and the use of secondary packaging quickly fell, although it was ordinarily less than 1 percent of all packaging. The main controversy has been around primary packaging, which represents the bulk of packaging materials.

The Duales System Deutschland (DSD)

To comply with the legislation, 400 companies set up the *Duales System Deutschland* (DSD) to try to meet the government's quotas for recycling the different packaging types. The DSD signs contracts with three groups. First, it licenses the use of its "Green Dot" symbol, as shown in Figure 5.1 below, to packaging producers, who put the logo

Figure 5.1
German Green Dot Symbol

on their packaging. Secondly, it contracts with private waste haulers and municipal waste collectors to collect packaging bearing the Green Dot. Thirdly, it contracts with industry organizations that guarantee the waste will be recycled.[3]

Material is collected from consumers and retailers and sent to secondary materials firms for processing. As much as possible, the material is reused to make new packaging, which will follow the cycle again. However, not all material can be reused. Some is used for producing products. An example of this is making park benches out of soda bottles. Other products will be used as sources of energy. Still other products may be exported. This product flow is shown in Figure 5.2.

Green Dot Program

The Green Dot program is the major focus of the DSD. Their logo is called a "Green Dot," even though on most packaging, it is printed in black ink.

In order for a container to be accepted by a participating waste hauler, the packaging must carry the Green Dot of the DSD. In order for a company's packaging to bear the Green Dot, the company must pay a licensing fee to the DSD. The cost for the Green Dot depends on a number of factors. The fees are based on the "producer pays" principle, and take into account sorting and recycling costs for the various packaging materials. For example, the costs for plastic are much higher than those for glass, because of the increased sorting and recycling costs of plastic. In addition, the volume of the product is taken into account.[4]

Figure 5.2
Flow of Materials in the DSD

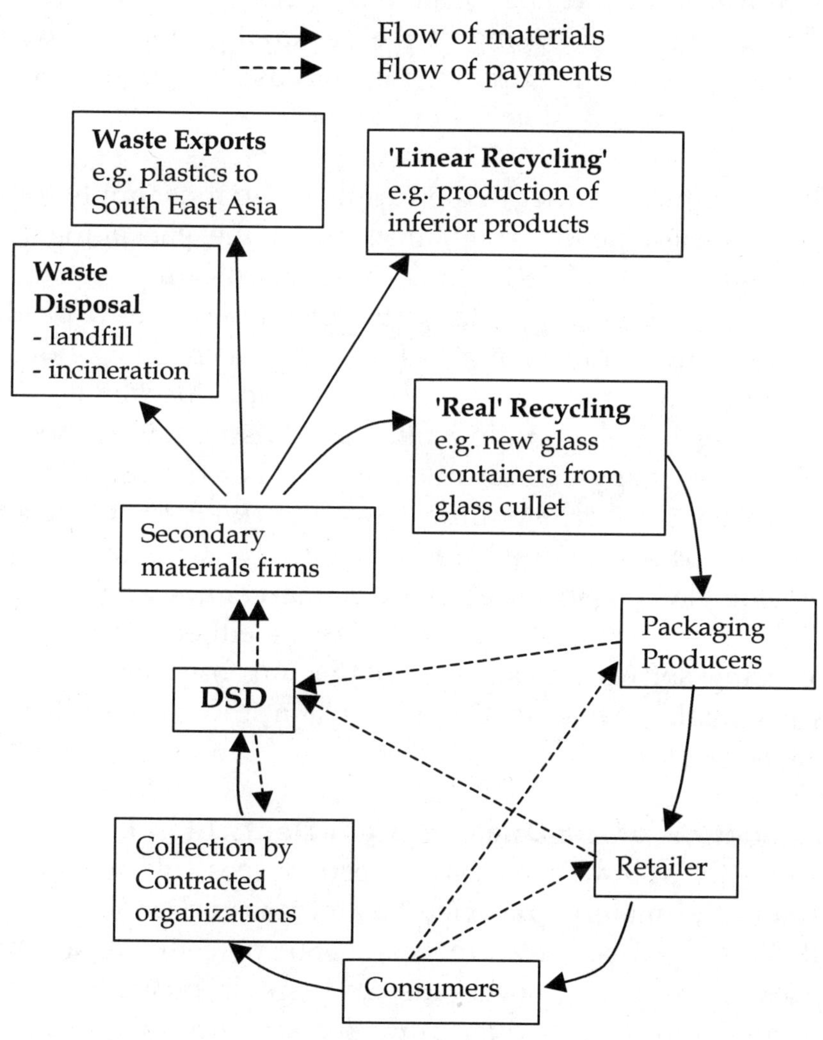

The "producer pays" principle dictates that the company responsible for an environmental situation should pay for the cost of the cleanup. In the case of packaging, the producer of the packaging should be responsible for the cost of recycling or disposing of the packaging. Obviously, the consumer will indirectly pay for this cost, but ultimately, the producer is responsible for the cost.

Exactly which partner in the supply chain must apply for the Green Dot depends upon the type of the packaging. For packaging that is filled with a consumer product such as, for example, a cottage cheese container, the product manufacturer must apply for the Green Dot, not the container manufacturer. For service packaging, like shopping bags, wrapping film, and disposable dishes, the packaging manufacturer must apply. For products that are imported, the German importer or the exporter, if located in the European economic area, must apply. Countries located outside the European economic area cannot apply for the Green Dot. Applications may be made either in paper form, or using software designed for the purpose, both of which are available by contacting the DSD at the address listed in the Appendix C.[5]

Evaluation of the Duales System Deutschland

The DSD's initiatives have been successful, in that the amount of material recycled has increased dramatically. By 1997, Germans were recycling 86 percent of all sales packaging from households and small businesses.[6] By comparison, many American cities are currently struggling to reach established targets of 25 percent .[7]

A criticism of the Green Dot system is that it is too expensive. Proponents of the original legislation claimed that recycling would be less expensive than sending products to the landfill. Thus far, recycling has not proven to be cheaper than landfilling. The system ends up costing less than $20 per person.[8] While this amount may seem high to some critics, it is much less expensive than many recycling programs in the U.S., which have achieved far lower rates of recycling.[9] The startup of the DSD was plagued by a number of problems which may strike any company attempting to set up a new reverse logistics system, although the scope and scale of this system are much greater than most reverse logistics systems will be.

All three of the partner groups that contract with the DSD experienced problems. The license fees collected by the packaging producers were initially too low to cover the cost of collecting and recycling the packaging. Also, many firms were slow to pay their agreed-upon licensing fees. As a result, the DSD narrowly avoided bankruptcy in 1993. Afterward, it revised the licensing fees to reflect the cost of recycling each type of waste, and increased pressure on companies to license the Green Dot and make prompt payments.[10]

The contracts with the waste haulers were more expensive than anticipated. In two years, the DSD had to sign an agreement with a company in each city in Germany to handle the collection. The waste haulers believed that finalizing agreements in all cities was more important than trying to keep costs down. Another factor which led to higher collection costs was Germany's inexperience with curbside recycling collection.   Prior to the packaging ordinance, recycling was done primarily through drop-off locations.   The final factor was that the DSD paid for collection on a per-ton basis. This gave the waste haulers no incentive to keep out products that did not belong in the system, which may have accounted for 20 to 40 percent of early collections.   As a part of its reorganizing after near bankruptcy in 1993, the DSD gave the waste haulers a per capita ceiling on their revenues for collection.   This gave waste haulers an incentive to reduce the amount of extraneous material entering the system.[11]

The DSD's hardest problems with may have been the companies it contracted to recycle the collected material.

In 1993, projections were for 300 million tons of recycled plastic. Actual results were 400 million tons. Processors were not prepared for such large quantities, and the market for recycled plastics was not yet ready to buy such large quantities.   When the recyclers accept plastic, they are obligated to ensure that the plastic is eventually recycled; that is, made into new products. Ideally, the material should all stay within Germany. With supply exceeding demand, prices of recycled plastic dropped dramatically, and large quantities were sent to other European countries, because companies desiring the materials could buy it much more cheaply from German sources than local sources.   This caused great problems for the recycling programs of other European countries, and has led to discussions about how much one country's policies should be allowed to impact the recycling efforts of other countries. The other result of the excess supply is that large quantities of plastic supposedly destined for reuse ended up in landfills in southeast Asia, as companies found it cheaper to export the problem than to find a use for it in Germany.[12]

The most interesting response to the flood of material created by the DSD has been the European Union's Packaging Directive, adopted in 1994. Perhaps for the first time anywhere, regulations attempt to restrict the amount of recycling. Under the directive, countries are to recover at least 50 percent of their packaging material, but no more than 75 percent. If countries want to recover a higher level of packaging, they must demonstrate that they have sufficient recycling capacity to handle their own materials.[13]

The other major complaint about the Green Dot program is that it is, in fact, a license for a company to create as much waste as desired. There is no incentive for a reduction in waste production. Also, there is no incentive for companies to explore reusable packaging. For this reason, some local governing bodies resisted the implementation of the program.[14]   Despite this criticism, the program has succeeded in reducing the amount of packaging waste Germans create. From 1991 to 1995, the amount of recycled packaging used by German consumers decreased by nearly 11 percent, while disposable packaging used in the U.S. increased 13 percent over the same period.[15]

In May of 1998, another German company, Landbell, began offering the same collection services as the DSD, but at a less expensive price. Unlike the DSD, the consumer does not separate each of the products to be recycled. The consumer must only separate paper and paper-based materials. All other household waste is collected and sorted centrally to remove metals, batteries, minerals, and glass. The remaining waste, including plastic, is dried, compressed into bales, and incinerated with energy recovery. The DSD is challenging the court ruling which allowed Landbell to begin service, on the grounds that it does not comply with Germany's 64 percent recycling target for plastics.[16]

5.2 Transport Packaging

Despite the fact that transport packaging is a significant portion of the total packaging used globally every year, it does not receive much attention in the U.S. The amount of transport packaging used every year, and its disposal, were discussed in the section on transport packaging, in Chapter 4.

In Germany, one of the goals of the Packaging Ordinance dealt specifically with transport packaging. Under the law, companies must take back all transport packaging that comes with its products, either for reuse or recycling. Therefore, all transport packaging must be coded for recycling. Figure 5.3 shows paving stones sitting on a pallet, wrapped in plastic which is coded for recycling.

Impact of Legislation on Transport Packaging
The immediate impact of this restriction was a rapid decrease in the amount of disposable transport packaging being used, and increased interest in returnable and reusable packaging. Demand for corrugated material in Europe is expected to decline by 10 percent over the next five years.[17]

**Figure 5.3
Transport packaging coded for
recycling in Germany**

The *Arbeitsgemeinschaft Verpackung und Umwelt*, (AGVU, literally translated as the "Working Party on Packaging and the Environment") is an association of German firms in retail, consumer goods, the packaging industry, and the recycling industry. Bringing together members of all parts of the supply chain, the AGVU's mission is to study the effect of Germany's closed-loop economy in the field of packaging, with a strong focus on the environment.

The AGVU commissioned the *Gesellschaft für Verpackungsmarktforschung* (GVM), a market survey agency for the packaging industry, to collect information from the government and other sources about Germany's use of packaging, in a wide variety of areas.

Although the members of the AGVU may not be enthusiastic about the Packaging Ordinance, the survey they commissioned demonstrates that the law has had a significant effect, as the following statistics demonstrate.

From 1991 to 1996, the annual usage of styrofoam transport packaging fell 36 percent, from 31 thousand metric tons to 23 metric tons. Over the same period, the use of pallets as transport packaging was reduced by 14.5 percent, from 385 thousand metric tons to 329 thousand metric tons. When other wooden packaging types are included (crates, wire spools, etc.), total use of wooden transport packaging decreased by 25.6 percent over the same period. Steel transport packaging decreased by 11.3 percent, corrugated transport packaging usage decreased by 8.8 percent, total paper transport packaging decreased by 7.6 percent, and plastic transport packaging decreased by 7.4 percent.

5.3 Product Take-Back

Throughout Europe, laws are increasingly forcing manufacturers to take responsibility for proper disposal. In some cases, the legislation mandates that manufacturers must be willing to take back products from consumers after the products have reached the end of their lives.

German Take-Back Laws

The most restrictive take-back laws in place are found in Europe, and the first and broadest of the European take-back laws are from Germany.

The new Basic Law of Waste Management was passed in 1993. Its goal was to move Germany closer to achieving a closed-loop economy, in which all production will be reused or recycled, with a minimal amount of production eventually being in the landfill. One of the key changes in this law is its definition of waste. The original Waste Management Act, passed in 1986, was only concerned with the regulation of waste for disposal. In the new act, the government is also concerned with waste for recovery. This change was needed in order to bring German law into agreement with European laws.

The major change brought about by the new legislation is that branches of industry, such as the electronic industry or the auto industry, are obligated to accept responsibility for their own products. This is a major step in moving from a throw away mentality toward preventing, reducing, recovering, and recycling waste. The ambitious goal of the law is to collect all production waste and used products, and forward them for recovery or recycling.

The ordinance for the auto industry came into force on April 1, 1998. Now, all automakers are required to take back old cars free of charge for up to twelve years after their initial registration. The ordinance for the electronics industry is not as far along, and it currently exists only in draft form.

In the future, ordinances will also be enacted regarding the proper disposal of printed paper, electrical appliances, batteries, and construction rubble.[18] Currently, standardized take-back laws are in development across the European Union. Although these policies are not yet implemented, work is under way to draft common policies for all member countries.

In different countries, the policies are different, but the major areas of concern can be broken into:

- White goods: refrigerators, freezers, heating equipment, water boilers, washing machines, dishwashers, and kitchen equipment
- Brown goods: sound equipment, televisions, photocopiers, and fax machines
- Computers
- Automobiles
- Batteries

Electronics Waste

Although studies have shown little environmental danger from disposal of electronics products, the issue of how electronic products should be disposed of, and how much recycling should be done of them is a widely discussed issue, especially in Europe. This is despite the fact that another study done by a British consulting firm reports that electronics products account for less than one percent of all solid waste in Britain.[19]

The European Commission presented a set of proposed laws on collection and recycling of electrical and electronic wastes in April 1998.[20] The final law will set targets for *Waste Electrical and Electronics Equipment* (WEEE), to increase recycling, to reduce hazardous substances, and to make sure waste left over after recycling is properly disposed.[21] The standards will cover a wide variety of electronic products, including cellular phones, games, toys, household appliances, and office machinery.[22]

The European Union (EU) proposal would apply the producer responsibility principle for the take-back proposals, meaning that manufacturers would be responsible for financing the cost of the system. This proposal upset member countries, which wanted to have flexibility in determining how these systems should be funded.[23] Manufacturers, while unhappy about having to pay for the recycling, would prefer uniform costs throughout Europe. Varying pricing throughout Europe could lead to distortions in competition.[24] Retailers are upset that the law requires them to accept goods from consumers making new purchases. Although manufacturers would have the ultimate responsibility for recycling, retailers would have to collect, store, and forward the products.[25]

While the EU begins consideration of a European-wide directive on electronics waste, many European countries are considering their own policies on electronics waste. For example, Norway has announced plans to require producers and importers of electronic equipment (EE) to take back discarded EE products and waste materials. Half a dozen trade organizations agreed to collect 80 percent of Norway's EE waste within five years. The agreement, which is scheduled to go into effect on July 1, 1999, covers discarded white goods, personal computers, telephones, cables, electronic and industrial electric goods. Following the polluter pays principle, the system will be financed by a recycling charge levied on new electronic products. The Netherlands and Switzerland are currently formulating similar policies, and the EU is formulating a directive on EE waste.[26]

In the Dutch electronics take-back system, the cost of collecting and recycling the products is to be paid from a levy placed on all new equipment sold. Although the fees will help recycle the new equipment currently being sold, it will not be adequate to cover the cost of all existing equipment—which will also need to be recycled. One issue to be resolved is to find out whether or not the government will financially support the organization's recycling of old equipment. Another issue to be resolved is related to the fact that 80 percent of the electronics sold in the Netherlands comes from one company. Other manufacturers resist the idea of paying for a system when the majority of the benefits will go to one firm.[27]

A study by a U.K. electronics recycling program found that television sets are currently too expensive to be recycled. The study concluded that "television sets are better stored in people's attics and garages until the process and the market for recovered materials is fully developed." Older sets contain more wood that is less valuable than the plastic used in newer sets. Also, cathode ray technology is not advanced enough to make television sets very good recycling candidates.[28]

Appliance Take-Back

In the Netherlands, companies that import electrical appliances have been ordered to set up take-back systems to collect and recycle used brown and white goods. After seven years of discussions on voluntary approaches failed to produce a workable result, the environment minister decided to force the companies to set up and run such a system. Under the system, companies will have to set up a system to collect and recycle the products. The system will be paid for by levies placed on products at the point-of-sale, which is not expected to significantly increase prices to consumers.[29]

Auto Recycling

According to an EU study, every year 9 million vehicles are junked in the EU, creating 9 million tons of waste. Draft legislation is under consideration; such legislation would mandate that by 2015, 90 percent of all vehicles must be recovered. Under the draft legislation, the last owner of a vehicle would receive a certificate of destruction from an authorized dismantler. If the vehicle has negative value, the manufacturer must reimburse the owner for this cost.[30] This last condition is a particular source of unhappiness for automakers.[31]

Automakers are also unhappy about targets the EU has set for non-recyclable auto content. The EU law has set targets for reducing the proportion of car waste that is landfilled — to five percent by the year 2015. Automakers say this will actually hurt the environment, because it will force them to use more metal, which is more easily recyclable, but heavier than other available materials.[32]

Unlike the automakers, the auto dismantlers of Europe welcome the law. They believe it will be an improvement over the system of voluntary agreements in place across Europe, which differ between countries.[33]

In the UK, a voluntary recycling system has been set up by representatives from vehicle manufacturers and dismantlers, and the metals, plastics, and rubber industries. A consortium representing all theses industries will manage the process. Unlike the schemes in the Netherlands and in Germany, the reclamation is to be governed by market forces. There will be no levy on new vehicles sold, and there will not be cost-free take-backs of old vehicles, as in Germany. To increase recyclability, manufacturers have committed to make new vehicles 90 percent recyclable by 2002. Unlike the EU scheme, the UK goals allow the inclusion of "energy recovery facilities," in which materials like plastic, rubber and fiber—which are difficult to recycle, and/or are combustible—can be burned.[34]

In the Netherlands, a national system of government-certified recycling centers has been established. At the time of purchase, the buyer pays a fee of 250 guilders (approximately $60.00). Once the car has been dismantled, the dismantler sends an official notification to Auto Recycling Nederland, the organizing body, who then disburses the fee. The ongoing operation of the system is also financed by a national auto ownership tax. Once a car is registered, the owner pays an annual fee until an official document can be produced proving the vehicle has been sold or dismantled.[35]

Although a number of EU countries have created voluntary take-back programs for automakers, Sweden was the first to announce a mandatory take-back requirement. The automakers must establish a network of dismantlers, retailers, and wholesalers to ensure that cars returned for recycling are properly dismantled. The manufacturers must accept the vehicles for recycling without charge, except in the case of older vehicles, when a fee may be charged if the cost of recycling will exceed the value of the materials reclaimed.[36]

Battery Collection

The EU is also considering a directive that would require 75 percent recovery of all kinds of batteries, and phase out all mercury and cadmium batteries by 2008. Such a directive is highly controversial, and could lead to significantly higher prices for toy, electronics, photographic, and tool and small appliance makers, as there is no current non-cadmium technology that will run small hand tools and some other electronics.[37]

In April 1997, the German Government adopted regulations regarding the recovery and disposal of used batteries. Manufacturers will be required to take back, recycle, and dispose of returned pollutant-containing batteries free of charge. This includes car batteries. Six months after it goes into effect, all other batteries will also be included. It is expected that this legislation will create the impetus for a unified battery return system.[38]

In Germany, legislation has also been drafted which would outlaw the sale of any appliances from which hazardous batteries cannot be easily removed.[39]

In Sweden, a 90 percent collection rate target for used nickel-cadmium batteries is in place. Because the industry did not meet this target in 1995, in 1997, the Swedish government put local officials in charge of collecting the batteries, while the industry must still pay for the collection.[40]

5.4 European Conclusions

Throughout Europe, there is a strong trend toward producer product take-back. In some countries, some industries are under voluntary take-back programs, in which the government and industry have agreed to targets that the industry will attempt to meet. If these targets are not met, these industries may find themselves under mandatory targets.

For a reverse logistician, different challenges are present in different countries, in different industries. In some industries, the government does the collecting, such as, in the Swedish battery industry. In some cases, a network of facilities is organized and run by the industry, like the Swedish auto industry, for example. In other cases, companies are left to create their own centers.

The money to pay for these systems comes in as many varieties as the systems themselves. In some cases, the industry must bear the cost, in some cases the user pays for disposal at the time of the product's purchase, and yet in other cases there are combinations of payment arrangements.

Chapter 6: Industry Snapshots

In this chapter, reverse logistics practices in selected industries will be examined. The industries included in this chapter were selected because each contains elements that can be translated to firms in other business sectors.

6.1 Publishing Industry

The publishing industry is currently struggling with record-breaking returns of unsold copies, a steady decline in adult trade sales, and a compressed shelf life for new titles. Reverse logistics is now more important in the publishing industry than ever. At many firms, good reverse logistics policies and practices represent the difference between profitability and seas of red ink.

Historical Roots

The book supply chain suffers from some problems that date back to business practices developed 70 years ago. During the Great Depression, booksellers could not afford to buy as many books (to sell) as the publishers wanted. To enable retailers to stock more books, publishers began the practice of permitting retailers to send back any books they were unable to sell. Retailers were then able to carry many more titles, thus greatly increasing the selection available to the buying public. Since then, publisher-retailer relationships have followed this model.[1]

This relationship requires the publisher to bear the risk for the books that fail to sell. Publishers encourage retailers to buy large quantities. Retailers know that any books they cannot sell can be returned for full credit, so there is zero or little cost to the retailer for ordering more copies than can be sold. This arrangement is very costly for publishers. Each return costs the publisher 25 cents in transportation, and many books are destroyed that typically cost between $2 to $2.50 to print.

Over time, publishers have come to accept these costs as a cost of doing business. Over the last few years, however, changes in the publishing industry have led to record return rates and losses for publishers.

In addition to buying directly from publishers, retailers also buy books through wholesalers. Retailers generally prefer to buy from the publisher, because these distributors generally charge higher prices. Lately, retailers in most channels have started gaining more power. In the book supply chain, book distributors have policies in place that limit their return risk, while publishers take the brunt of the return risk. Some publishers will now only sell directly to retailers making a minimum level of annual purchases. Smaller retailers must buy from a wholesaler. Because wholesalers often do not carry a publisher's entire catalog, this is a serious concern for smaller retailers.[2]

Return Problem Symptoms

Sales of hardcover and paperback adult trade books fell by 5.3 percent between 1995 and 1996, to 459 million copies.

This was the second consecutive year that the drop in sales was greater than five percent. From 1996 to 1997, total revenues declined 3.4 percent.[3] One publisher canceled over 100 planned titles in 1997. During this period of decreasing sales, the average rate of return in that category hovered at about 35 percent of copies shipped to booksellers. Additionally, the product life cycle of book sales has decreased.[4]

According to the American Association of Publishers,[5] return rates for 1996 were as follows:

Table 6.1
Returns as a Percentage of Gross Revenues, 1996

	Average	Range
Adult Hardcover	35.1%	23%-38%
Adult Paperback	25.6%	22%-29%
Juvenile Hardcover	18.7%	13%-23%
Juvenile Paperback	19.8%	13%-23%
Mass Market Paperback	43.5%	37%-51%

The most closely watched category of books is the adult hardcovers. These are usually written by well-known authors that typically dominate best-seller lists. Sales figures for these books can have an impact on the stock price of a publisher. Return rates in this category have dramatically increased, and are such an area of concern that the publishing industry is devoting more resources to solve the problem.[6]

Root Causes

Several causes exist for the return problems in the publishing industry:

- Rapidly growing retail square footage requires more books
- Chains' size has led to larger print runs
- Chains generally have higher return rates
- Competition for likely best-sellers has increased advances given to authors, requiring bigger runs to recoup initial investments
- Profusion of books in print means more competition
- Shorter shelf lives
- Flat total sales growth for books overall
- Computer models exert downward pressure on shelf life
- "Jam the channel"
- Inventory policy changes—JIT
- Unclear channel position—integration

The largest single factor has been the growth of the large chain stores. More than 80 superstores opened during the period 1990-1996, with the top four chains (Crown Book Corp., Books-A-Million Inc., Barnes & Noble and Border's Group, Inc.) opening more than 190 stores in 1996. In 1996 alone, there was a 20 percent increase in the amount of retail bookselling space.[7] The amount of space is expected to grow by another 14 million square feet over the next few years.[8]

Retail floor space grew exponentially as chains such as Barnes & Noble and Borders placed large footprint book superstores all over the United States. Warehouse clubs, such as Sam's and Price Costco, also devoted large spaces to heavily discounted books. At the same time that retail space was expanding, consumer spending on books was slowly growing. While there is now much more retail space devoted to selling books, there is not that much more demand by consumers for books.

At the same time that retail floor space was growing in the book industry, the customer base for books and many other manufactured items was consolidating. Giant retailers, such as Wal-Mart and Kmart, have become a larger percentage of a publisher's business. As superstores add more space, they are grabbing a larger and larger piece of the book market, and these gains are coming at the expense of independent bookstores. Figure 6.1 shows the independent bookstores' portion of the market. As late as 1991, independent bookstores held the largest share of the market, with more than 32 percent, only to decline to less than 20 percent by 1996.[9] The number of books sold by discount stores rose by more than 21 percent in 1996.[10]

The growth of superstores has given bookstores a more powerful position in negotiations with publishers. In order to secure a prominent display in the superstores, publishers must be able to supply large quantities of books. However, after two weeks in a prominent display, a book may be relegated to a shelf in the back of the store. The large display may include a stack of 100 or more books. When the

Figure 6.1
Change in Independent Booksellers Market Share

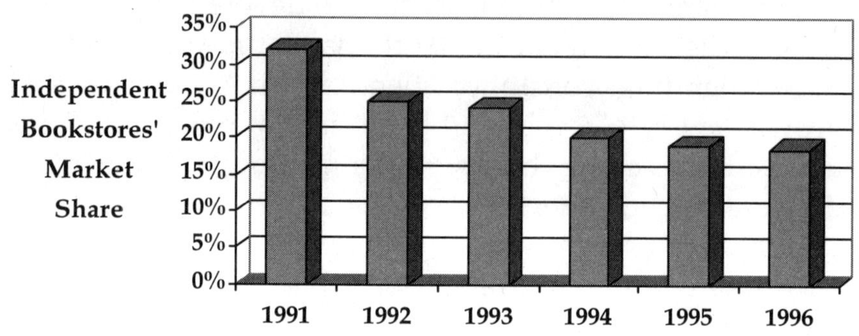

large display is no longer needed, most of those excess copies are sent back to the publisher for credit.[11]

The result is that the increase in sales at superstores is a mixed blessing for publishers. Independent bookstores sell more than 80 percent of the books they order, while superstores sell less than 70 percent, and discount stores sell about 60 percent. Because independents tend to order a smaller variety, less of each title, and push the books they have in inventory, their return rates are generally lower than those of the superstores. Additionally, independent booksellers have less power in the channel. The growth away from independents to the powerful superstores has exacerbated the return problem in the book industry.

If a store has a large inventory of slow-selling books, it will often want to mark them down to be able to sell them, rather than return them. However, publishers generally believe that marking books down diminishes the value of the product in the customers' eyes. In the publishing industry, it is believed that if the customers expect that all books will eventually be marked down, they will postpone purchases of new profitable books.

Another problem that publishers have is the explosion of titles in the publishing industry. In 1947, when *Books in Print* began collecting data, 357 publishers printed 85,000 titles. By 1997, there were more than 1.3 million books in print, published by more than 49,000 publishing houses in the U.S.[12] Yet, the number of customers that read books has not grown at the same rate. In 1997, sales slid 3.4 percent.[13] The result is that book superstores often dictate reverse logistics policies to the publishers.

Making the problem more financially potent, publishers are gambling on high profile authors with huge advance payments.   To compete for these high-profile authors, advances paid are running into the millions of dollars. To justify such high advances, publishers plan initial print runs at least large enough to cover the author's advance. This leads to more copies of the book to distribute, which adds to the problems described above. Unfortunately, many of the authors do not return what the publisher expected.

The Internet and On-Demand Publishing
At one point, some people believed that bookstores would eventually become obsolete.   The vision existed that publishers would sell books directly to people who would

download books from the Internet. Interestingly, one of the areas where the Internet has had the largest impact on commerce is in the publishing industry. *Amazon.com* has shown through its success that there are many people who will use the newest communication form, the Internet, to order books, one of the oldest forms of communication.

In the future, the Internet's greatest impact may be on sales of out-of-print books. Because of traditional printing technology, publishers must print a large number of copies of a book at one time to keep costs down. If it is unlikely that future sales will guarantee at least this many sales, the publisher will not print another run of the title, and the book will become obsolete.

Technology is making it possible to print very small runs as small as one copy for a reasonably low cost. The cost is still much higher than the per-copy cost of printing several thousand, but low enough to still be affordable.

Using this technology, a customer can go to a website and request a copy of an out-of-print book. The book will be printed and shipped to the customer the next day. Ingram Books, the nation's largest book wholesaler, has a Lightning Print division devoted to this very business. Their website, www.lightningprint.com, offered 125 titles in July of 1998, and the company has made plans to offer 10,000 titles by the end of 1998. The cost (to the publishers) of making a title available is relatively low, as the company is letting publishers set up titles risk-free at no up-front cost. The publisher will collect royalties on books that would have never been sold otherwise.

6.2 Computer / Electronic Industry

One executive said to the research team, "We're in an industry with 60-day product life cycles and 90-day warranties. Of course customers are going to bring products back." The product life cycle of a computer is extremely short when compared to other consumer durable goods, such as automobiles or large appliances. In a business where returns can lower profits by as much as 25 percent, reverse logistics is a serious business.

According to the Gartner Group, the used PC business was between $2-3 billion in 1996. Approximately 25 million obsolete PCs became ready for remanufacture or disposal this past year. Given a population of approximately 260 million in the United States, that is just under one obsolete computer per person. A study completed by Carnegie Mellon University, estimates that approximately 325 million personal computers will have become obsolete in the United States in the 20-year period between 1985 and 2005. Out of that number, it is estimated that 55 million will be placed in landfills and 143 million will be recycled.[14] This large number of obsolete computers means that the short life cycle in the electronics industry is a serious problem, and that there are many opportunities to reuse and create some value out of a nearly omnipresent asset.

Figure 6.2
Problems with Computer Lifecycles

From the Wall Street Journal -
Permission, Cartoon Features Syndicate

For many retailers that sell computers and electronics, the percentage of returns is high. Manufacturers have begun to put caps on the amount of product they are willing to take back. These caps are part of a continuous struggle in the channel. Because computers are a complex product, return percentages are high. Consumers do not understand how to operate them and are quick to return the product when it may not be defective. Some categories, like CD-ROM drives, have had return rates of 25 to 40 percent in the past because they were complicated to install and difficult to operate.

Printer returns, on the other hand, have moved down to between four and eight percent because they have become an appliance. The consumer can simply unpack the printer, plug it in and start using it.

For one computer manufacturer, failure to manage the return rates well severely damaged its profitability, and eventually, its ability to go public. This retail channel firm allowed its return rates to get out of line compared with the rest of the industry. Return rates for PC firms included in this research are generally below 10 percent. However, this particular company allowed its return rates to exceed 17 percent.

Many computer manufacturers have put caps on their returns, and allow only a certain percentage of sales to be returned. These policies have been known to fail when a powerful retailer tests them and exceeds the cap. It is difficult to manage powerful retailers such as Wal-Mart or Target, when manufacturers are dependent on those retailers for a growing percentage of their sales. Some firms started out with an aggressive cap percentage have since eased the percentage to accommodate their retail customers.

One way to minimize the return chain is by building to order. This allows manufacturers to postpone final transformation of the product until the end of the channel, and configure the exact computer that the customer wants. With postponement, the channel holds very little inventory. This is in sharp contrast to the rest of the industry, which typically will have 30-60 days of inventory pre-sold into the channel.

The manufacturers/retailers that sell directly to the customer and build to order have significantly lower return rates than the rest of the industry. These firms have return rates around five percent, about half of what the rest of the industry experiences. One executive interviewed said, "we send out a million computers. Pretty soon, most of them come back." The build-to-order model, combined with direct sales, eliminates this problem.

The direct manufacturers/retailers interviewed for this research find that the bulk of their returns is due to quality problems. Most of the returns for manufacturers that sell computers through the traditional reseller's channel are marketing returns, where the computers did not sell and came through the channel to the manufacturer. Direct manufacturers/retailers have a clear advantage over traditional competitors because of the minimization of returns. Additionally, direct sellers believe that most of their users are higher up the technology curve and therefore are less likely to ship back the non-defective defective machines that stream back to all computer manufacturers. One traditional firm interviewed indicated that nearly half of its bad quality returns were actually working models. However, each firm's business model dictates how reverse logistics works.

One computer manufacturer that at the time of this writing was just beginning to move to a build-to-order manufacturing model, flooded the market with inexpensive computers. Soon after it jammed the channel with these machines, it changed the rules on returns and price protection. The company decided to not allow any open or closed box returns, and to limit defective returns to one percent of resellers purchases from the previous quarter. Non-defective defectives were to be returned to the customer for full price. An executive with one of this manufacturer's competitors talked about this development with the research team. "What an interesting concept. First you flood the channel with excess inventory, then you announce that returns are prohibited." This policy is clearly a tentative step in the direction of making return policies more conservative, and places more responsibility downstream in the supply chain. It remains to be seen if this and other similar initiatives will be successful.

Computer manufacturers have developed rebate programs to incent retailers to reduce returns. For example, one manufacturer gives retailers a one percent rebate for return rates between four and seven percent, and up to two and a half percent if returns are less than one percent for that particular retailer.

Some manufacturers have contracted with remanufacturing specialists to develop solutions to this problem.   These companies will work with manufacturers, evaluate the root causes of returns, excess, and obsolete machines, and develop methods to control costs and return rates. These specialist firms test, recondition, repair, repack, and then resell the machines. One firm, for example, includes a special manufacturer's warranty. At the same time, the third party can act as the service center for the manufacturer. Some manufacturers have also hired third parties to perform warranty repair and other service work for retailers that do not have their own service capabilities. These programs have led to lower returns.

Businesses have begun to learn that the largest portion of their profits is derived from the early stages of the product life cycle. This knowledge makes proper product disposition even more important. One electronics company interviewed for this research said that it made 140 percent of its profits during the first four months of the product's life. This statement means that in the latter portion of the product life cycle, where sales begin to dwindle, profits are actually negative for this particular electronics firm. This situation is not unusual. In the electronics industry, as in many other industries, product life cycle continues to contract. Retailers realize that they have to get a product through the supply chain quickly, get that product on their shelves, and then move it off the floor before it becomes unprofitable. The backward portion of the supply chain then becomes a priority rather than an afterthought.

Toner Cartridges

Toner cartridges for laser printers have become a major area of remanufacturing. Initially, toner cartridges were difficult to recycle, but manufacturers have since learned how to make them easier to disassemble and refurbish.

One of the first well-known programs to be billed as reverse logistics was Canon's Clean Earth Campaign. With each new cartridge sold, customers received a mailing label with which to send their old cartridge to Canon, at no cost. The program was very successful, and Canon received many cartridges back for remanufacturing.

A new toner cartridge can easily sell for approximately $100. A remanufactured cartridge will sell for 25 percent less. The cost to remanufacture a cartridge is very low, making remanufacturing of toner cartridges a very profitable business.

Unfortunately for manufacturers, cartridges are so easy to remanufacture that virtually anyone can start a business in their garage remanufacturing cartridges. Companies on the Internet are advertising kits that contain the necessary tools and materials needed to start a home-based cartridge remanufacturing business.

The toner cartridge remanufacturing business has grown to include 12,000 remanufacturers, employing 42,000 workers, selling nearly $1 billion annually.[15]

Given the profits involved, it is not surprising that many companies now advertise that they will pay $10 or more for used toner cartridges. The original cartridge manufacturers find themselves competing against these remanufacturers to get their own cartridges back.   One toner cartridge manufacturer perceived this to be such a large problem that it introduced a plan to prevent its customers from selling their used cartridges to other companies. Instead of receiving a rebate for sending the cartridge back to the company, customers receive a "prebate" at the time of purchase. In return, the customer promises that the cartridge would not be reused, refilled, or remanufactured. Environmentalists and the remanufacturing industry quickly rose up in arms against the proposal.[16]

Software Industry

In the software industry, distributors are attempting to cut down retailers' returns by implementing just-in-time delivery. However, retailers generally overestimate demand because there is not much incentive for them to forecast carefully. Software manufacturers want the product on the retailer shelves, and often agree to stuff the channel. The cost of a box of software is low compared to the price. In one extreme example, a software manufacturer contracted with a third party to destroy 50 million copies of one software product. While this particular manufacturer would have preferred to not produce an excess of 50 million, the company believes that it is better to guess higher than lower. Because of these kinds of practices, return rates in the software industry recently hovered around 20 percent. Additionally, releasing more software titles forces returns, because the product life cycles of those titles are contracting.

Because their risk is low, some retailers will accept software purchased elsewhere. Other retailers, such as Sears, are trying to reduce returns and improve inventory turnover by reexamining channel relationships. Some of these retailers have begun setting up 30-day return policies.

6.3 Automotive Industry

The auto industry is one of the largest industries in the world, dealing with the most expensive consumer goods. Therefore, it is not surprising that reverse logistics issues have long been a source of consideration. In the auto industry, there are three primary areas in which reverse logistics plays a significant role:

- Salvage of parts and materials from end-of-life vehicles
- Remanufacturing of used parts
- Stock-balancing returns of new parts from dealers

This section concentrates only on the North American auto industry. The European auto industry has been dealt with in some depth in Chapter 5. The auto industry also makes extensive use of returnable containers. Returnable containers in general have been discussed in Chapter 4.

Auto Disassembly

When a vehicle reaches the end of its life, it eventually ends up at an auto salvage yard or auto dismantler. There, an assessment is made of the components of the vehicle. Any parts or components that are in working order that can be sold as is, are removed and sold. Other components, such as engines, alternators, starters, and transmissions may be in fairly good condition overall, but need some refurbishing or remanufacturing before they can be sold to a customer. Once all reusable parts have been removed from the vehicle, its materials are reclaimed through crushing or shredding.

Shredded metals will generally be reclaimed, but the remaining material, known as fluff, cannot be recycled.

Every year, automotive recyclers handle more than 10 million vehicles. Their efforts supply more than 37 percent of the nation's ferrous scrap for the scrap-processing industry.[17] However, roughly 25 percent (by weight) of the material in a car is not recycled in the United States.

Approximately 35 percent of the nonmetal material left after shredding a car is plastic. To reduce the amount of landfilled plastic, firms are trying a number of alternatives. One part of the problem is that cars are made of so many different types of plastics. Identifying each type of plastic is difficult. Automakers are trying to reduce the number of types used, and to label the parts for easier separation after disassembly. Ford, for example, reduced the number of grades of plastic that it specifies from 150 to 20.[18]

To increase the recyclability of cars, the big three automakers in the U.S. have joined together to form the Vehicle Recycling Development Center (VRDC). At the VRDC, they are trying to learn how to build vehicles to be disassembled more easily. They are investigating one of the newest trends in engineering, Design For Disassembly (DFD).

Unlike other environmental initiatives for manufacturing, DFD offers the possibility of many unintended positive effects. Disassembly of a product is made easier by reducing the number of parts, rationalizing the materials, and snap-fitting components instead of chemical bonds or screws.

These objectives fit in well with other current manufacturing strategies: global sourcing, design for manufacture, concurrent engineering, and total quality.[19]

In one example of DFD from the computer industry, Siemens Nixdorf's PC41 introduced in 1993 contains 29 parts, versus 87 in its 1987 predecessor. The new machine can be assembled in 7 minutes and disassembled in 4. The old machine took 33 minutes to assemble and 18 to disassemble.[20]

Although there are many automobile salvage operators in the U.S., automotive material suppliers are in general agreement that the largest obstacle to increased automotive recycling is the lack of a nationwide network of sophisticated automotive dismantlers.[21]

Use of Recycled Materials

To close the recycling loop, automakers would like to be able to use recycled products in their vehicles. However, parts made out of recycled materials are not yet widely available. When they are available, they may cost more than parts made of virgin materials. However, Ford discovered in one case, that once all the costs of using a particular part are considered, a 100 percent recycled part was actually cheaper to use. Unfortunately, this is not the case in every instance. Because automakers believe that consumers will not pay extra for a vehicle made with recycled parts, additional usage of recycled materials will depend on the rate at which their costs can be brought down.[22]

Automakers are making progress in this area, however. Chrysler, for example, recently announced their consideration of a program to take material from pop bottles and use it to make large panels for the body of a car. The car will be very lightweight, and at this point, able to meet all U.S. safety requirements, except side impact collision standards. However, it would initially be targeted for areas of growth in low-priced vehicles, like China and India.[23]

Remanufacturing

The auto industry may be the industry with the longest history of making use of old products. The remanufacturing of auto parts was boosted by the shortage of new parts during World War II, but the recycling of auto parts has been taking place in the industry for over 70 years.

According to the Auto Parts Remanufacturers Association (APRA), the remanufactured auto parts market is estimated at $34 billion, annually. The APRA also estimates that there are 12,000 remanufacturing firms involved in the auto parts industry. Although there are many firms involved, there are also many large firms in the industry. One company remanufactures more than 4 million alternators, starters, and water pumps every year.[24] In total, 90 to 95 percent of all starters and alternators sold for replacement are remanufactured.

Automakers want to maintain a closed-loop system with their parts. When a vehicle needs a new transmission, it is their hope that the consumer will bring the car to a dealer, who will replace the old transmission with a remanufactured one. The dealer will send the old transmission (now called a transmission "core") to the automaker for remanufacturing. In this way, the automaker will maintain a stable supply of transmission cores.

Unfortunately for the automakers, there is a lot of leakage from this closed system. To prevent this, the dealers must pay a deposit in the form of a core charge when they receive a remanufactured part. When the automaker receives the transmission core from the dealer, the dealer's core charge is refunded.

Despite these efforts, many parts leave the system. Partly, this leakage occurs because many car owners take their vehicles to auto repair shops outside the automaker's system. The core will then go to the repair shop's supplier.

Another source of leakage can be the dealers themselves. Despite the fact that the dealer may have paid a core charge for a part, they may still be willing to sell it to another remanufacturer. Many of the third party remanufacturer companies make regular milk runs, during which they stop at dealers and other repair shops and offer to pay cash for any cores. Many dealers will look at a pile of greasy cores sitting in the corner, and decide they would rather take the cash than deal with the hassle of sending the cores back to the automaker.

All automakers interviewed recognize that this leakage is problematic, and are working on ways to improve their reverse logistics processes to eliminate this problem. For example, Ford has begun using a single carrier to handle all of its dealer core returns. Ford dealers can call one 800 number for all questions and issues related to core returns.[25]

The auto parts remanufacturing business can be difficult. All of the typical problems of reverse logistics are present: varying flows of different products, and many products without packaging. This is not to mention the fact that the products themselves are often coated with grease. However, some remanufacturers have begun using bar coding systems to track incoming products and the progress of these products as they move through their reverse logistics flow.[26]

Dealers Parts Returns

In addition to collecting cores from dealers, automakers also must collect new, unused products and defective products from dealers.

Although the automakers wish that they could determine the amount and type of parts the dealers will maintain in their inventories, and how many, they cannot. Dealers, as independent businesses, make their own determinations as to which parts they stock. Each year, new car models are introduced requiring new parts. At the same time, fewer older cars are on the road, meaning parts for these vehicles no longer need to be kept in the dealers' inventories. Because dealers have a finite amount of space in which to store parts, they need to remove the older parts from their inventories to make room for the newer parts.

Many auto dealers are family-based businesses with limited supplies of capital to invest in inventories. They often have less than state-of-the-art inventory management capabilities. It is in the best interest of the parts supplier to clean out their inventories, reduce credit-line constraints, and improve customer satisfaction.

The research team interviewed automakers about parts returns. To help the dealers maintain an inventory of current parts, automakers allow dealers to return a limited amount of parts for full credit every year. The exact amount is different for each manufacturer, but the calculation is the same: the value of the returned parts must be no more than some percent (typically four to six percent) of the value of the new parts the dealer has purchased during the year. To deter dealers from abusing the system, some automakers allow dealers to receive a check for some percentage of their unused return allowance.

The truck that delivers new parts to the dealership typically picks up the returned part, and the part will go back to the parts distribution center. What happens when the parts arrive at the parts distribution center depends on the automaker. At some facilities, the parts are inspected to make sure that the right part is in the box, and the part is put back on the shelf. At other facilities, the parts are put into new packaging so that the dealer who next receives the part will be unable to tell that the part has been returned. At one distribution center, a separate 100,000 square foot facility is used for this purpose. At other distribution centers, separate repackaging contractors are used.

At one automobile manufacturer, dealer resistance to receiving returned parts is extremely high. If a dealer receives a part and suspects that it has previously been returned, the dealer will immediately send the part back. Therefore, rather than even try to repackage the parts, all returned parts under a rather high dollar amount are destroyed and sent to the landfill.

6.4 Retail Industry

The retail industry, under great competitive pressure, has used return policies as a competitive weapon. The greater the pressure, the more innovative the solutions. Within the retail industry, it appears that necessity, indeed, is the mother of invention.

Grocery retailers were the first to begin to focus serious attention on the problem of returns and to develop reverse logistics innovations. Their profit margins are so slim that good return management is critical. Grocery retailers first developed innovations such as reclamation centers. Reclamation centers, in turn, led to the establishment of centralized return centers. As covered in detail in Chapter 2, centralizing returns has led to significant benefits for most firms that have implemented them.

Over the last several years, retailers have consolidated. Now more than ever, large retail chains are the rule. These large retailers have more power in the supply chain than retailers did a few years ago. In general, the large retailers are much more powerful than the manufacturers. Few manufacturers can dictate policy to large retailers such as Wal-Mart or Kmart. If a manufacturer will not accept returns, it is unlikely that the large retailer will accept those terms easily. In some exceptional cases, retailers will make allowances for a manufacturer's products that they believe are not replaceable with similar products.

Returns reduce the profitability of retailers marginally more than manufacturers. Returns reduce the profitability of retailers by 4.3 percent. The average amount that returns reduce the profitability among manufacturers is slightly less, at 3.80 percent.

Survey respondents were asked how they disposition returns. On average, retailers use a centralized return facility to handle returns much more frequently than manufacturers. Retailers are also found to be more likely to sell returns to a broker or similar entity. They were less likely to remanufacture or refurbish than manufacturers — which would seem logical given that manufacturers are better at manufacturing than retailers. Manufacturers are significantly more likely to recycle or landfill returned material than retailers. It appears that retailers are further advanced than manufacturers when it comes to asset recovery programs. For other disposition options, such as resold as is, repackaging, or donation; retailers' responses were quite similar to manufacturers.

In Table 6.1, a comparison of disposition options between retailers and manufacturers is presented.

Table 6.2
Comparison of Disposition Options between Retailers and Manufacturers

Disposition	Retailers	Mfgs.
Sent to central processing facility	29.2%	17.7%
Resold as is	21.4%	23.5%
Repackaged and sold as new	20.5%	20.0%
Remanufactured/Refurbished	19.9%	26.7%
Sold to broker	16.8%	10.1%
Sold at outlet store	14.5%	12.8%
Recycled	14.1%	22.3%
Landfill	13.6%	23.8%
Donated	10.6%	11.8%

Technology

It is clear both from the interviews and the survey instrument that retailers have made larger investments in technology to improve their reverse logistics systems. In fact, manufacturers lag behind retailers in almost every technology category. This difference between manufacturers and retailers does not appear to exist in all facets of an operation.

Nearly twice as many retailers as manufacturers included in the research implemented automated material handling equipment. Retailers are also more likely to use bar codes, computerized return tracking, computerized returns entry, electronic data interchange (EDI), and radio frequency (RF) technology to enhance their reverse logistics management. A comparison of reverse logistics technology adoption is presented below in Table 6.2.

Table 6.3
Comparison of Technologies Utilized to Assist Reverse Logistics Processing By Retail and Manufacturing Segments

Technology	Retailers	Mfgs.
Automated material handling equipment	31.1%	16.1%
Bar codes	63.3%	48.7%
Computerized return tracking	60.0%	40.2%
Computerized returns entry at most downstream point in supply chain	32.2%	19.1%
Electronic data interchange (EDI)	31.1%	29.2%
Radio frequency (RF)	36.7%	24.6%

6.5 Conclusions

Reverse logistics practices vary based on industry and channel position. Industries where returns are a larger portion of operational cost tend to have better reverse logistics systems and processes in place. In the book industry, where great change in the industry structure has occurred in the last few years, returns are a major determinant of profitability. In the computer industry where life cycles are nearly as short as grocery life cycles, the speedy handling and disposition of returns is now recognized as a critical strategic variable. Successful retailers understand that managing reverse logistics effectively will have a positive impact on their bottom line. Industries that have not had to spend much time and energy addressing return issues are now trying to make major improvements. Now, more than ever, reverse logistics is seen as being important.

Chapter 7: Future Trends and Conclusions

Recognition of Reverse Logistics

It is clear that in the future, more firms will lavish considerable attention on reverse logistics. Many firms have only become aware of the importance of reverse logistics relatively recently, and have yet to realize the strategic importance that reverse logistics can play.

To reduce the cost of reverse logistics, in the future, firms will need to focus on improving several aspects of their reverse logistics flows:

- Improved gatekeeping technology
- Partial returns credit
- Earlier disposition decisions
- Faster processing / shorter cycle times
- Better data management

One of the easiest ways to reduce the cost of a reverse logistics flow is to reduce the volume of products it is asked to carry. There are two aspects to this. First, products that do not belong in the flow should be prevented from entering. Secondly, once products have entered the flow, they should be dispositioned as quickly as possible.

7.1 Reducing the Reverse Logistics Flow

To reduce the flow of products into the reverse logistics system, there are a number of promising new technologies that can be used to make sure that every product that enters the reverse logistics flow is one that should be in the system.

Product Life Cycle Management

Good reverse logistics management can be considered part of a larger concept called product life cycle management. In the future, it is likely that leading edge companies will begin to emphasize total product life cycle management. The product life cycle management concept means that the firm provides the appropriate logistics and marketing support based, at least in part, on where the product is in its life cycle.

The core of the product life cycle concept is that all products have a finite life and move through various stages. Typically, a product life cycle curve is divided into four distinct phases during its life as a live product. Those four phases are introduction, growth, maturity, and decline. Product volume increases through the introductory and growth phases. As a product moves through the life cycle to maturity, sales level off and begin to decrease. In the declining stage, sales drop and profits derived from the product diminish.

Products in various stages of the life cycle require different types of management and support. Logistics management needed in the introduction phase is much different than the support requirements when the product is at maturity. Additionally, the supply chain management necessary at the end of a product's life varies from other stages.

The product life cycle is not uniform across products and industries. It is a theoretical device that can be useful. However, it is difficult to identify where a real product is in the life cycle once it moves past the introductory and growth stages. The firm has to look for demand turning points. These can only be seen if the firm clearly understands past history and the marketplace.

Unfortunately, the focus for the marketing and logistics organizations at many firms is only on the early and middle portions of the product life cycle. The mission of the logistics and marketing organizations is much clearer early in the life cycle. Product roll-out, volume build, and the support associated with these portions of the life cycle, are their primary concerns. Sometimes, it is difficult for a company to admit that a product is at the end of its life. Decisions are postponed because the organization believes that a little more life can be breathed into a product. The sales and marketing organizations may attempt to conceal a sales decline. They may believe that a decline in the sales of a product means that they are not performing their job properly.

As the product approaches the end of its life, the cost of holding inventory increases. Inventory carrying costs consist of expenses such as the cost of money, insurance, taxes, shrinkage, warehousing, and obsolescence. As the product moves through the life cycle and reaches the end of its selling life, obsolescence costs increase from very low to 100 percent. Warehousing costs associated with the product will also continue to accumulate. This means that a firm cannot correctly use only one inventory carrying cost across the total life of the product. In Figure 7.1, an inventory carrying cost scenario is presented.

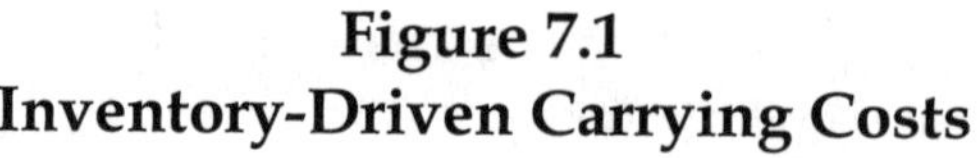

Figure 7.1
Inventory-Driven Carrying Costs

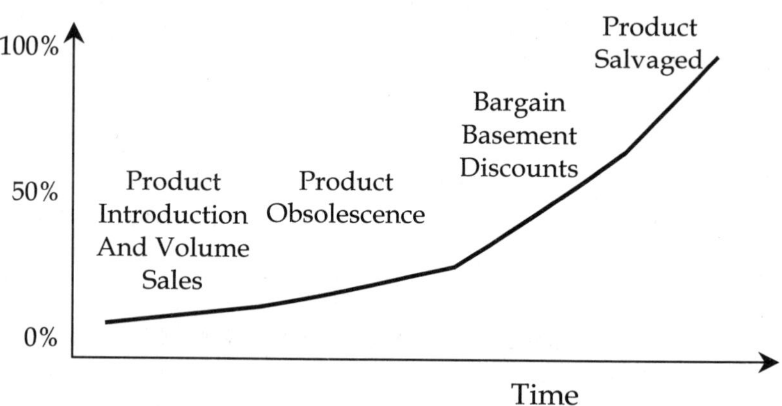

It is as important to manage products well at the end of their life as it is in the beginning. As can be seen from the figure, it may be more important to manage inventory well at the end of a product's life than at the beginning.

At the end of a product's life, it is likely that it will enter the reverse logistics flow. Good reverse logistics is a critical piece of product life cycle management. As the life cycle moves past volume sales, the firm needs to begin to clear the channel through the utilization of good reverse logistics practices. Plans must be made for the end of product life, as well as thinking about the other stages of the life cycle. If a firm can plan many of the management elements around the end of a product's life, instead of merely reacting late to obsolete inventories, the total profit derived from a product will be greater.

Information Systems

In order to handle reverse logistics better, firms will need to improve their reverse logistics information systems. As explained in Chapter 2, the reverse logistics environment is different enough from the forward channel that information systems developed for the forward channel do not generally work well for reverse logistics.

Most return processes are paper-intensive. Automation of those processes is difficult because reverse logistics processes have so many exceptions. Reverse logistics is typically a boundary-spanning process between firms or business units of the same company. Developing systems that have to work across boundaries add additional complexity to the problem. To work well, a reverse logistics information system has to be flexible.

Information systems should include detailed information programs about important reverse logistics measurements, such as store compliance, return rates, recovery rates, and returns inventory turnover. Some of the systems for controlling returns will obviously require significantly expanded and improved information systems.

Even if such systems do not materialize, firms will develop better reverse logistics information systems in the future.

For many companies, current information systems do not allow them to monitor the status of their returns.

Additionally, useful tools such as radio frequency (RF) are helpful. New innovations such as two-dimensional bar codes and radio frequency identification license plates (RFID) may soon be commonplace.

Gatekeeping Technology

In order to improve gatekeeping, front-line employees need good information about which products to allow into the reverse logistics flow. Accomplishing this task is not easy. It is made more difficult because many front-line employees who are making gatekeeping decisions are often inexperienced exmployees working at or near minimum wage. Retailers are loath to incur significant costs training these employees, because employee turnover tends to be high. More training would certainly improve gatekeeping, but using a significant amount of training on an ongoing basis would not be cost-effective.

If it is not feasible to provide the gatekeepers with a high level of training, many manufacturers have sought to "bulletproof" the returns process by providing materials for the employees to follow when taking a return. Such materials would let the employee know what products can be returned, for how long after purchase, and what parts should be included with the product. For example, in the case of a VCR, the employee should make sure that the VCR, the remote control, and the cables are all present, and that the remote matches the VCR. In an effort to improve retailers' returns processing, a well-meaning manufacturer may put together a binder outlining each of the parts that should be present for each one of its products.

Unfortunately, these carefully thought-out materials are often obsolete within months or weeks of printing due to product changes.   The most significant problem these materials face is the fact that they are very paper-intensive. Imagine the scene at a typical returns desk: a half a dozen or more manufacturers have sent their own three-ring binders to a store's returns desk, followed by periodical updates. Soon, the returns desk is swamped in materials, and no employee knows where it all is, nor which materials are obsolete or which are current. Rather than trying to sort their way through the morass of procedures and policies, employees just ignore them.

Web-Based Gatekeeping

As described in Chapter 2, some retailers are investing in gatekeeping systems. One solution that seems promising is an internet or intranet web page that guides the employee through the returns process for each product.   When a customer returns a product, the employee scans the UPC bar code on the product.   The computer system asks the manufacturer's system for the returns procedure for the particular product.   A web page appears, which steps the employee through the returns process for that product.

Using the VCR as an example again, after the employee scans in the VCR's UPC code, a web page appears with a picture and short description of the product.   The employee is asked if the VCR matches the product shown on the screen. If not, the employee could choose the proper product from a pull-down menu. The page would include pictures of any accessories that should be present. In addition to the VCR itself, there would be a picture of the remote control and a picture of the cables.

One problem retailers and manufacturers have repeatedly mentioned in the current study is that employees do not appreciate the importance of having the accessories present. Some manufacturers will not allow a product to be returned if certain key accessories are missing, or will only provide partial credit. A retailer may have given a customer a full refund for a returned VCR, but only receive a credit for 50 percent of the cost from the vendor because the remote control was not present.

For this reason, manufacturers and retailers alike have said that they wish it were possible to make the employee aware of how much it will cost the store if certain components are missing.

For example, next to the pictures of the VCR's remote control and cords, the web page could list the cost to the retailer of not having each of these items. This would make the employee aware of the importance of having each of these components. Some components may have a greater cost than others. A lost remote control may cost the retailer $5, whereas a lost cable may have zero cost.

Electronic Data Interchange
Another important technology is Electronic Data Interchange (EDI). While this technology is not new, most of the firms included in the research had not yet fully adopted EDI. While many processes within an organization may be automated, it hard to marshal the resources to implement all of the EDI transaction sets that a firm might wish to have. Obviously, for most companies the reverse flow would not be among the first to adopt and implement.

A description of EDI is listed in Appendix D. This appendix contains a detailed discussion of the 180 returns transaction set.

POS Registration
In some cases, manufacturers are willing to accept customer returns for a limited period of time after the initial purchase. If a retailer attempts to return a product to the manufacturer after this period has expired, the manufacturer will not give the retailer credit. In this type of situation, the retailer needs to know exactly when the product was purchased.

A technology that can provide this information is point of sale (POS) registration. In a POS registration system, the retailer scans the product's serial number at the time of sale. The retailer electronically sends the serial number and the sale date to the manufacturer. The manufacturer keeps on file the serial number, the sale date, and the name of the store that sold the product. When a customer tries to return a product at a later date, the retailer phones the manufacturer to learn if the product is within the warranty period.

If a web-based returns system like the one above is implemented, this function could be included in the web page. After the employee scans the product UPC, a web page appears that instructs the employee to scan in the product's serial number. The web page accesses the database, and tells the employee whether the product was purchased at the retailer's store, and if it is possible to return the product.

Such a system nearly eliminates products being improperly returned. In the case described, the manufacturer pays the retailer fifty cents for every product registered. Clearly, this system comes at some cost to the manufacturer. In addition to the cost of registering the products, the manufacturer must also bear the considerable cost of developing and implementing such a system.

Despite the cost of such a system, for some products, the benefits can be great. The benefits will be greatest for high value products with short life cycles. When a product has a short life cycle, the customer may have a greater incentive to try to return a product beyond the authorized warranty returns period. When a new version of a product is released, many customers want to return the old version for a new version. If too much time has passed since the purchase for the manufacturer to authorize the return, some customers will try to abuse the returns system. The more frequently a new version of the product comes out, the greater customers' incentive to abuse the system. Also, the higher the value of the product, the greater customers' incentive to abuse the system.

Fighting Returns Abuse

POS registration is but one example of technology that could become more widely used in the future, as retailers and manufacturers join forces to fight returns abuse. Clearly, both manufacturers and retailers suffer when consumers fraudulently return products. Although manufacturers and retailers often are on opposite sides in many issues, fighting returns abuse is an issue on which both sides can agree to work.

RFID

Keeping track of where reverse logistics products are and where they are going can be time consuming. Many products do not have their original packaging, or the packaging may be damaged. In this case, it is very difficult to use RF scanners to track the movement of products through the reverse logistics flow.

Radio Frequency Identification (RFID) is a relatively new technology that may prove beneficial in these situations. Typical methods for identifying products are passive. In order to know if a particular product is present, the only way to find out is to go out into the warehouse and look to see if it is there. RFID, in contrast, is a more active form of identification. A very small, very low powered radio transmitter is installed in each product, broadcasting a very faint signal. Despite its small size, an RFID "tag" contains a battery that can send out a signal for years. The signal is strong enough, that it can be picked up by receivers in a warehouse. Each product can send a different signal. You could build 10 million computers, and install an RFID "tag" in each one, and each one could have a different signal.

There are two ways to use RFID: passively and actively. In passive RFID, a "sentry" at the entrance to the warehouse records the identification of each product as it enters the warehouse. Then, its ID is also registered when it leaves the building. Any items that have entered, but not left the building must still be in the warehouse. In active RFID, receivers are placed throughout the warehouse. To find out if a particular item is in the warehouse, the receivers listen to see if the product's signal is being received. If it is being received by more than one receiver, using triangulation, it is possible to determine where in the warehouse the product is.

Using RFID to assist in the management of returned computers might be a good option. Placing an RFID tag on the machine at the time of manufacturing would take away the errors in the paper chain and assist in the life cycle management of the computers.

One firm interviewed in the research has developed an RFID system that simultaneously reads multiple passive tags contained in various cartons on a pallet. Its system scans multiple items without those items needing to be unpacked. For certain products, an RFID tag could be placed on the circuit board or embedded in the plastic case of the computer.

RFID has the potential to aid reverse logistics operations in a number of ways. As mentioned, it may be helpful in keeping track of products in the warehouse. The other way it may be beneficial is in gatekeeping. RF tags may be used in recording the ID of products when they are sold, and this information can be useful in determining which products to accept for return.

Two-Dimensional Bar Coding

Two-dimensional bar coding is another technology that holds promise for reverse logistics operations. Two-dimensional bar coding schemes, such as PDF417 or Maxi-Trac, allow the user to embed much more information in a bar code than one-dimensional systems such as UPC. One-dimensional systems contain a number or code that must be translated by the computer and matched with information already had inside the machine. With two-dimensional bar code systems, the bar code can contain not only a code, but also a description and other text, even as long as Abraham Lincoln's Gettysburg Address.

Because reverse logistics transactions and processes are often exception-driven, information required to update the computer may not be able to fit within the limitations of one-dimensional bar codes. This limitation of one-dimensional bar code schemes could mean that for reverse logistics applications, new technologies—such as RFID or two-dimensional bar codes—will become the rule rather than the exception.

Data Collection

Using improved product tracking technologies like those just described, and with improved information systems for monitoring the flow of these products, firms will have much more information available about their reverse logistics processes. With this much more information available, the challenge will be for firms to effectively use it to their advantage.

In the preceding chapters, a number of examples have been given of firms that have used reverse logistics systems to watch for quality problems with their products. By tracking the number of returns of a product, a firm can be aware of an unusually large number of returns, and take appropriate action.

Design for Reverse Logistics

Some firms have developed programs to design products to be more easily manufacturable, Design For Manufacture (DFM). Others have designed items to flow through the supply chain more efficiently, Design For Supply Chain Management (DFSCM). In the near future, perhaps more companies will begin to think about designing products for reverse logistics management. A suggested term for this idea is this Design For Reverse Logistics (DFRL).

In some cases, firms have designed reverse logistics capabilities into the product (for example, Nintendo's scannable, see-through window mentioned in Chapter 2). Most direct retailers include address labels so that customers can easily ship the product back. Some firms, such as Canon and Xerox, want to reclaim more value from used copier and printer cartridges, and incent consumers to return empty cartridges when they purchase new ones. DFRL is to design reverse logistics requirements into product and packaging. It is the integration of reverse logistics needs and environmental concern into the product and the reverse logistics chain.

Partial Refunds

Once a returns information system is in place, a firm can be more precise in determining the amount of credit a customer should receive. Instead of giving the customer a full refund, and absorbing the cost of the missing remote, the retailer may elect to only give the customer a partial refund, holding the customer responsible for the cost of the missing remote. The retailer may, in fact, go beyond this cost, and add an additional, punitive charge for the incomplete return.

Partial refunds offer a helpful middle ground, as opposed to forcing the employee to make a choice between accepting or refusing the product for a full refund. If the employee refuses to accept the product at all, and the customer has, in fact, lost the missing component, the customer may become irate. Allowing the employee the additional option of giving the customer a reduced refund may be a preferable option.

Partial refunds will give the customer a strong incentive to make sure that any returned products are, in fact, complete, which should also reduce the retailer's cost of returns.

Regardless of whether a retailer implements such an advanced system, improving the technology available to employees responsible for gatekeeping, will be an important area for reducing reverse logistics costs.

7.2 Managing Reverse Logistics Flow

Once products have been allowed into the reverse logistics system, companies must manage the flow of these products to minimize their net impact to the bottom line.

Standardization of Processes

One of the most common difficulties the research team observed with current reverse logistics systems is the lack of standardization of processes throughout an organization. If processes are not standardized, it is very difficult for people in an organization to communicate to each other how to handle reverse logistics problems.

Good reverse logistics processes begin at the retail store by simplifying returns policies and procedures. These simplified policies and procedures should translate into fewer labor hours dedicated to returns processing. Higher quality decisions should also result because of simplification.

As described earlier in this book, for a variety of reasons, reverse logistics often is a low priority area at many companies. In the future, companies will increasingly recognize the strategic importance of reverse logistics, which should result in more corporate resources being available for improvements to reverse logistics flows.

Many companies have discovered that the major benefits of ISO 9000 certification of their forward channel are derived from standardizing all of their processes. Although many firms may not elect to get their reverse logistics processes ISO certified, as more resources become available, many firms will appreciate the benefit of standardization.

Centralized Return Centers

As discussed in Chapter 2, research respondents who had implemented a Centralized Return Center (CRC), were in agreement that using a separate CRC offered many benefits. At a CRC, employees have a much larger volume of products to deal with than they would ever experience at a retail store. This allows employees to develop areas of expertise, which can greatly benefit the firm.

Many of the benefits cited appear to be due to the fact that this allowed the returns processing staff to focus solely on returns. Asking staff to "serve two masters" by having people responsible for both forward and reverse distribution, seems to work poorly. The benefits may not arise because of the CRC being physically separate from the forward distribution centers (DCs), but because of the separation of control.

In the future, firms will continue to benefit from separating the control of the forward and reverse channels. However, it would seem likely that firms will learn how to handle returns in a CRC that may be at the same physical location as a forward DC. This will allow firms to place their CRCs at the best geographical location, regardless of the presence or absence of a DC.

Third Parties

A major logistical trend of the 1980s and 1990s has been the recognition of logistics as a field in its own right. As firms have analyzed their core competencies, many have realized that they lack the expertise, skills, and experience to perform certain functions in-house. In the future, many firms may determine that reverse logistics falls into the category of activities which are best outsourced.

Firms will realize that efficiently handling the reverse flow and maximizing revenues from secondary markets are specialized skills. Many firms that thirty years ago would have never considered outsourcing their distribution, now find they can significantly reduce costs by using third party logistics providers. In the future, many of these same firms may come to similar conclusions about reverse logistics activities.

Secondary Markets

In the future, firms are likely to find themselves working much more closely with their partners in the secondary market. The current logistics paradigm of moving material from the plant to the finished goods distribution center, and then to the store, will expand to include the passing of this inventory along to the secondary market.

In the last 10 years, outlet stores have risen from a small portion of total sales to a much larger percentage. As described in Chapter 3, outlet stores were originally conceived to dispose of unsold or returned merchandise. As the outlet store segment grew, a larger quantity of material was needed to keep this new channel well stocked. In response, many firms began to produce products especially for the outlet store market.

It is difficult to predict what the future will hold for the secondary market. From all indications, it will continue to grow. As the secondary market grows, some manufacturers will take measures to increase their control over what happens to their products once they leave the "A" channel.

Web-Based Secondary Markets

A trend in the disposition of goods is the utilization of the worldwide web. While at the time of this writing utilization of the web is limited, it appears that in the future it will be an important mechanism for dispositioning from the reverse logistics flow.

The web provides a direct link to consumers. By eliminating several intermediate steps, a web-based supply chain can be more efficient than a typical channel. This efficiency can be very useful when selling product that has entered the reverse logistics flow.

Some product is sold to the secondary market electronically through a web auction. A web auction allows consumers to bid on products until a preset closing time. The web sites such as *www.onsale.com* or *www.surplusdirect.com* enable the liquidation of merchandise while maximizing revenues. Generally, these web sites offer a few pieces of an item and allow consumers to bid on them. Instead of selling these items at a fixed price, offering them via an auction mechanism means that customers will bid the price up to an equilibrium point where demand meets supply. Holding back some of the product encourages consumers to bid up the price of the limited number of items that are available for auction that day.

The web provides a quick feedback mechanism to inform the firm whether there is a profitable market for the product. If the product does not perform well at the web auction, then perhaps another disposition option should be selected.

A limiting aspect of web auctions is that only consumers with web access can use them. As usage of the Internet grows, usage of the web will grow and this problem will decrease. Currently, the largest portion of secondary market items available via the web is computer-related.

Zero Returns

As described in Chapter 2, a number of firms are experimenting with zero returns programs. In a zero returns program, the manufacturer never again takes possession of a product once it has been sold.   The retailer takes responsibility for dispositioning product in accordance with the manufacturer's stipulations. In return, the retailer receives a payment that is intended to reimburse him for the cost of the returned items and for dispositioning the product. By removing the need to handle the returns, the manufacturer expects to save enough costs to more than offset the increased payments to the retailer.

Under some zero returns programs, the store always receives a credit for a certain percentage of sales, no matter how high the return rate. If the credit is six percent, and actual returns are only two percent, the retailer is happy, because it still receives a six percent credit. When the opposite happens, the credit is set at two percent, and returns are six percent, the retailer loses. The idea behind the program is that the credit will be set high enough that it will exceed the average returns experienced by the retailer. However, given the power held by the large retailer chains, it can be difficult for the manufacturer to prevail against the retailer in this situation.

Unfortunately, given the lax controls that many firms keep over their returns, controls over zero returns programs are lacking. Some manufacturers accuse retailers of double dipping, taking payment from the manufacturer for destroying the product, and then quietly selling the product out the back door to a secondary market firm.

Reverse Logistics Strategies

Depending on the life cycle of a manufacturer's products, and the value of the products, firms will discover that different combinations of the above strategies will be needed to effectively and efficiently handle their returns.

For high-value products with short life cycles, like computers, video games, and camcorders, a POS system may be a very efficient way for retailers and manufacturers to reduce the costs of fraudulent returns. However, the cost of POS registration may make it difficult for many items to be managed in this manner. For example, the cost to track an individual low cost item, such as a pair of jeans, would probably prohibit using POS registration.

Process Improvement

In attempting to improve reverse logistics processes, a firm can move along several fronts. Suggested improvements described earlier in this book are listed in Table 7.1

Table 7.1
Key Reverse Logistics Management Elements

- Gatekeeping
- Compacting Disposition Cycle Time
- Reverse Logistics Information Systems
- Central Return Centers
- Zero Returns
- Remanufacture and Refurbishment
- Asset Recovery
- Negotiation
- Financial Management
- Outsourcing

7.3 Conclusions

While much of the world does not yet care much about the reverse flow of product, many firms have begun to realize that reverse logistics is an important and often strategic part of their business mission. Throughout the course of this research project, many examples of large bottom-line impact were identified. There is a lot of money being made and saved by bright managers who are focused on improving the reverse logistics processes of their companies. It is clear that, while sometimes derisively referred to as junk, much value can be reclaimed cost-effectively. While the efficient handling and disposition of returned product is unlikely to be the primary reason upon which a firm competes, it can make a competitive difference.

Appendix A: Letter/Copy of Survey

March 11, 1997

Good Morning:

The Center for Logistics Management at the University of Nevada is working on a research project to study reverse logistics trends and practices. We know that you probably receive many questionnaires, but we are asking you to complete one more.

Enclosed is a dollar as a token of our appreciation. Please complete the questionnaire and return it at your earliest convenience. If you are unable to complete the questionnaire, please just put it in the mail anyway. Your assistance is critical to a better understanding of reverse logistics trends and practices.

All responses will be kept strictly confidential. No one else but the research team will ever view any of the raw data. No company data will be identified. Should you have any questions, please call me at (702) 784-6814.

Thank you again for your kind assistance.

Sincerely,

Dale S. Rogers, Ph.D.
Director

THANK YOU

We sincerely appreciate your help in filling out this questionnaire. Your rapid response is critical to completing this research and will help develop a better understanding of Reverse Logistics in North America

Reverse Logistics Research

Conducted by:

**University of Nevada
Center for Logistics Management**

MAILING INSTRUCTIONS

This survey booklet is designed to be mailed as is. The "to" and "from" addresses are already on the back cover. Should you wish to seal it for confidentiality, use tape. No postage is necessary.

Dr. Dale S. Rogers
Center for Logistics Management (024)
College of Business Administration
University of Nevada, Reno
Reno, NV 89557-0016
(702) 784-4912
fax (702) 784-1773
e-mail: logis@unr.edu
web page: http://unr.edu/homepage/logis

For this questionnaire, we are defining Reverse Logistics as the <u>process</u> of moving goods from their typical destination to another point for the purpose of capturing value otherwise unavailable, or for the proper disposal of the product.

1. **How long is the life cycle of a typical product?**
   - ❏ Less than 3 months
   - ❏ More than 3 months to 6 months
   - ❏ More than 6 months to 12 months
   - ❏ More than 12 months to 18 months
   - ❏ More than 18 months to 2 years
   - ❏ More than 2 years to 3 years
   - ❏ More than 3 years to 5 years
   - ❏ More than 5 years

2. **On a scale of 1 to 7, with 1 representing very conservative return policies, and 7 representing very liberal return policies, how would you rate your policies regarding customer returns?**

Very <u>Conservative</u>						Very <u>Liberal</u>
1	2	3	4	5	6	7

3. **How, if at all, have your return policies changed in the past year?**

More <u>Conservative</u>						More <u>Liberal</u>
1	2	3	4	5	6	7

4. **What role do returns play in your company s strategy? Check all that apply.**
   - ❏ Clean channel
   - ❏ Protect margin
   - ❏ Competitive reasons
   - ❏ Recapture value
   - ❏ Recover assets
   - ❏ Legal disposal issues
   - ❏ Other, please specify____

5. **By what percentage do returns reduce your profitability?**

   ____________ %

6. **What, would you estimate, is the impact your returns have on your profits? (as a percentage of profits)**

   ____________ %

7. **What percentage of your total Logistics costs do your Reverse Logistics costs represent?**

   ____________ %

8. **Which of the following Reverse Logistics activities does your company perform either in-house or by utilizing a third party?**

	In-House	Third Party
Centralized collection center	❏	❏
Refurbishing	❏	❏
Remanufacturing	❏	❏
Outlet sales	❏	❏
Salvage	❏	❏

9. **Where in the supply chain are decisions made about what is to be done with a returned item? Additionally, is a third party used to perform any of this decision making?**

	In-House	Third Party
At retailer (or point of customer contact)	❏	❏
At regional distribution center	❏	❏
At national distribution center	❏	❏
At a returned goods processing center	❏	❏
Other, please specify		
_______________________________________	❏	❏

10. **Of the products that are returned by your customers, please estimate the percentage of goods represented by each of the following:**

Activities	Percentage
Donated	______ %
Recycled (materials reclaimed)	______ %
Remanufactured/Refurbished	______ %
Repackaged and sold as new	______ %
Resold as is	______ %
Sent to central processing facility	______ %
Sold at outlet store	______ %
Sold to broker	______ %
Land Fill	______ %
Other, please specify	
____________________________	______ %

11. **What barriers to successful Reverse Logistics Activities exist in your firm? Check all that apply.**
    - ❏ Company policies
    - ❏ Competitive issues
    - ❏ Financial resources
    - ❏ Importance of reverse logistics relative to other issues
    - ❏ Lack of systems
    - ❏ Legal issues
    - ❏ Management inattention
    - ❏ Personnel resources
    - ❏ Other, please specify ____

12. What hardware and software technologies do you have installed, or plan to install, to assist your returns handling?
❏ Automated material handling equipment
❏ Bar codes
❏ Computerized return tracking
❏ Computerized returns entry at most downstream point in supply chain
❏ Electronic data interchange (EDI)
❏ Radio frequency (RF)
❏ Other, please specify ____

13. How long is the returns processing cycle time for most of the products you handle?

❏ Less than 1 day	❏ More than 2 weeks to 1 month
❏ More than 1 day to 2 days	❏ More than 1 month to 2 months
❏ More than 2 days to 1 week	❏ More than 2 months
❏ More than 1 week to 2 weeks	

General Company Information:
14. What is your primary business?
❏ Building, Materials, Hardware, and Garden Supply
❏ General Merchandise
❏ Electronics and Computers
❏ Food
❏ Automotive
❏ Chemical
❏ Paper and Forest products
❏ Apparel and Accessory
❏ Furniture, Home Furnishings, and Equipment
❏ Drugs, Health & Beauty Aids
❏ Warehousing
❏ Trucking
❏ International logistics third party
❏ Other, please specify _________________________________

15. On a scale of 1 to 7, with 1 being very unimportant, and with 7 being very important, rate the importance to your customers of each the following in their decision to use you as their supplier:

Factor	Least Important						Most Important	
Cost reduction	1	2	3	4	5	6	7	N.A.
Price	1	2	3	4	5	6	7	N.A.
Quality of service	1	2	3	4	5	6	7	N.A
Return policies	1	2	3	4	5	6	7	N.A
Speed of delivery	1	2	3	4	5	6	7	N.A.
Variety of products	1	2	3	4	5	6	7	N.A.

16. What were the annual gross dollar sales of your business during the most recent fiscal year?

❏ $5 Million or Less	❏	Over $150 to $200 Million
❏ Over $5 to $10 Million	❏	Over $200 to $250 Million
❏ Over $10 to $50 Million	❏	Over $250 to $500 Million
❏ Over $50 to $100 Million	❏	Over $500 to $1 Billion
❏ Over $100 to $150 Million	❏	Over $1 Billion

17. How many people do you currently employ at this facility?

❏ 100 or fewer	❏	251 to 500
❏ 101 to 150	❏	501 to 1,000
❏ 151 to 250	❏	Over 1,000

18. In which of the following channel positions do you operate? Check all that apply.

- ❏ Manufacturer
- ❏ Wholesaler
- ❏ Retailer
- ❏ Service Provider (Please explain:____________________________________)

Name: ____

Title: ____
Parent
Company: ____
Division or Business
Unit: ____

Street Address: ____

City: ____________________________State: _________Zip:__________________

Telephone: ____

Fax:____

E-mail Address: ____

Do you wish to receive a copy of the survey results?
- ❏ Yes   ❏ No

Would you be willing to participate in a short telephone follow-up interview to this questionnaire?
- ❏ Yes   ❏ No

THANK YOU FOR YOUR PARTICIPATION.
THIS INFORMATION WILL BE HANDLED IN A CONFIDENTIAL MANNER.

Appendix B: Data Tabulation

The following are tabular and graphical representations of the data gathered from the surveys.

Question 1: How long is the life cycle of a typical product?

<3 months	10.13%
3-6 months	9.47%
6-12 months	13.51%
12-18 months	12.50%
18-24 months	11.82%
2-3 years	8.45%
3-5 years	8.45%
>5 years	25.67%

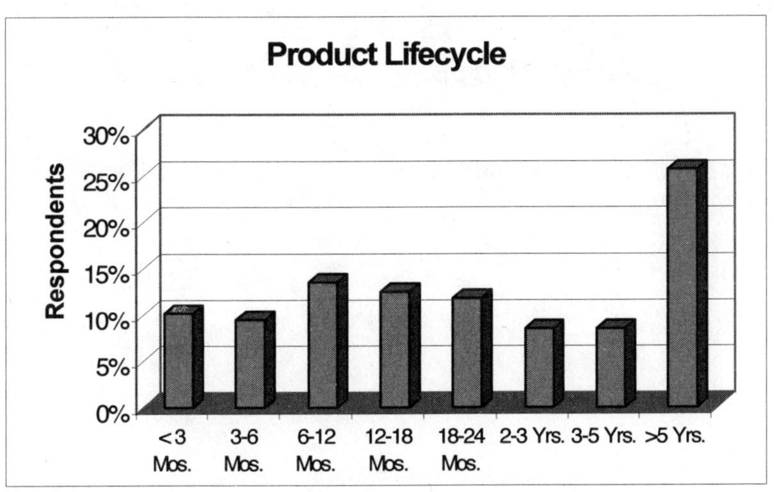

Question 2: On a scale of 1 to 7, with 1 representing very conservative return policies, and 7 representing very liberal return policies, how would you rate your policies, regarding customer returns? (1=very conservative, 7=more liberal).

Most of the participants showed a liberal return policy for their products.

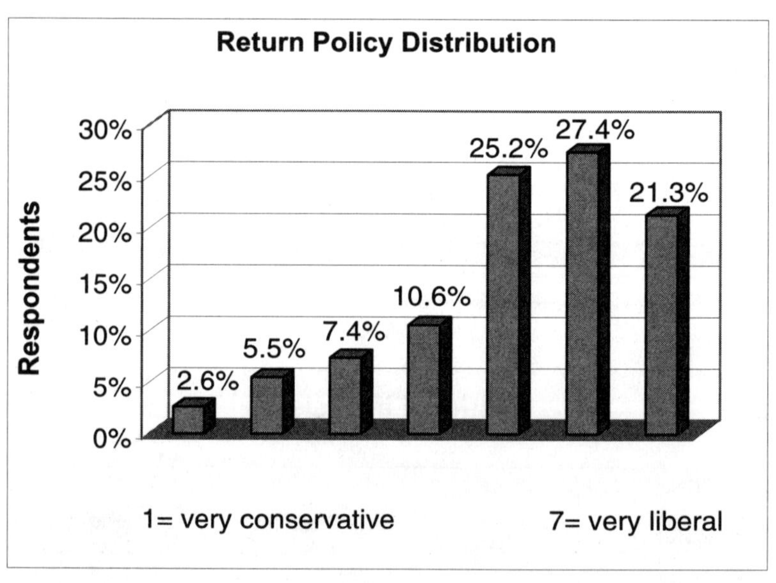

Question 3: How, if at all, have your return policies changed in the past year? (1=more conservative, 7=more liberal).

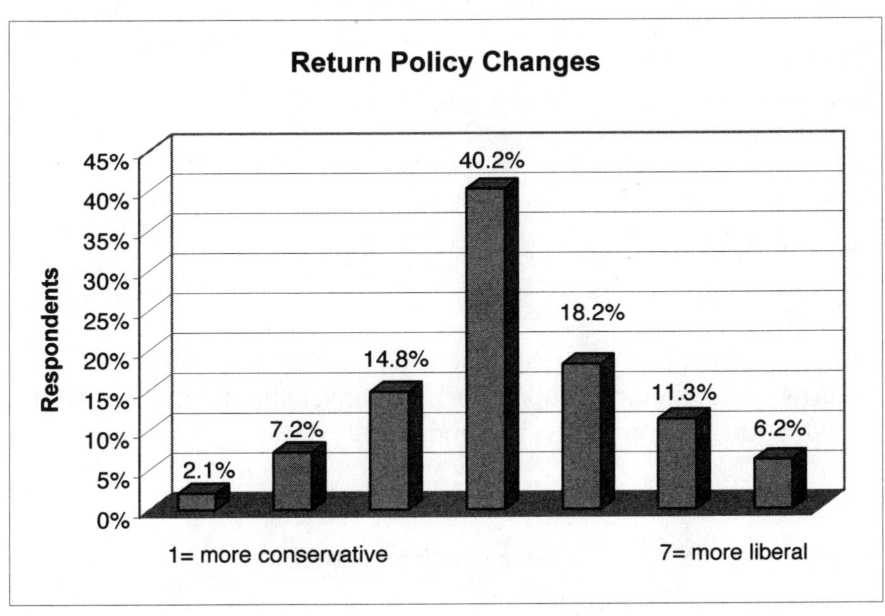

Question 4: What role do returns play in your company s strategy? (Check all that apply.)

<u>Returns role in company s strategy</u>	<u>Respondents</u>
Competitive reasons	64.9%
Clean channel	33.1%
Legal disposal issues	29.3%
Recapture value	28.6%
Recover assets	27.6%
Protect margin	18.3%

Also, 17 percent of the respondents specified other roles played by returns in their company's strategy, including: customer satisfaction, customer service, and quality.

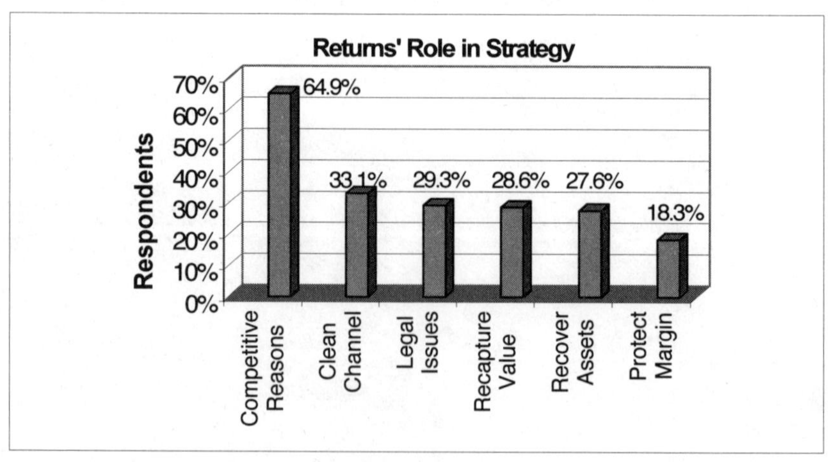

Question 5: By what percentage do returns reduce your profitability?

The average percentage by which profitability is reduced is 4.2 percent, with a standard deviation of 9.8.

Question 6: What, would you estimate, is the impact your returns have on your profits? (as a percentage of profits)

The average impact returns have on profits is 3.7 percent, with a standard deviation of 5.9.

Question 7: What percentage of your total Logistics costs do your Reverse Logistics costs represent?

The average impact returns have on profits is 3. 9 percent, with a standard deviation of 6.3.

Question 8: Which of the following Reverse Logistics activities does your company perform either in-house or by utilizing a third party?

	In-House	**Third Party**	**Both In-House and 3PL**
Centralized Collection Center	49.2%	15.1%	3.5%
Refurbishing	38.7%	9.3%	3.9%
Remanufacturing	29.9%	9.9%	0.6%
Outlet sales	28.6%	8.0%	2.9%
Salvage	45.9%	17.0%	5.2%

Question 9: Where in the supply chain are decisions made about what is to be done with a returned item? Additionally, is a third party used to perform any of this decision making?

	In-House	Third Party	Both In-House and 3PL
At retailer (or point of Customer contact)	44.4%	3.9%	1.3%
At regional distribution center	37.9%	4.2%	1.3%
At national distribution center	38.3%	1.3%	0.3%
At a returned goods processing center	27.6%	8.4%	1.3%

Question 10: Of the products that are returned by your customers, please estimate the percentage of goods represented by each of the following:

Activities	Average Percentage
Resold as is	17.6%
Remanufactured / Refurbished	15.5%
Recycled (Material Reclaimed)	14.7%
Landfill	13.9%
Repackaged and Sold as New	11.0%
Sent to Central Processing Facility	9.0%
Donated	6.8%
Sold to Broker	5.6%
Sold at Outlet Store	5.1%

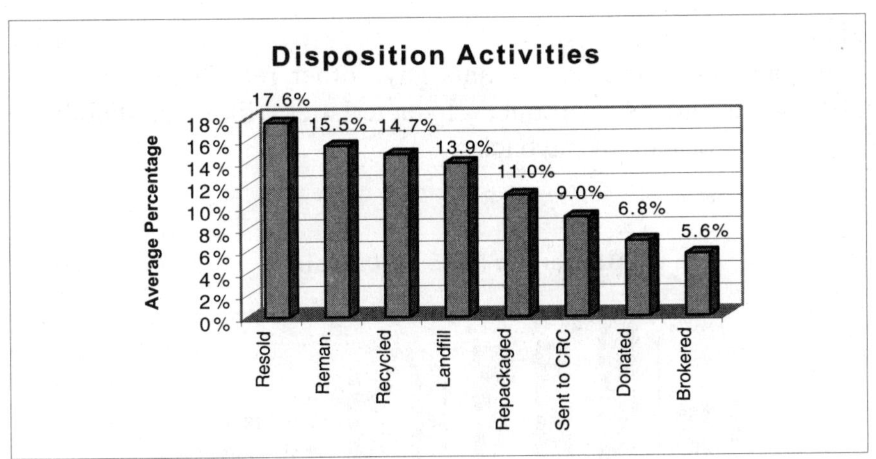

Question 11: What barriers to successful Reverse Logistics Activities exist in your firm? Check all that apply.

Barriers	Respondents
Importance of reverse logistics relative to other issues	39.9%
Company Policies	35.4%
Lack of systems	35.1%
Competitive issues	32.1%
Management inattention	27.3%
Personnel resources	19.3%
Financial resources	18.9%
Legal issues	14.1%

Other

Seven percent of the participants gave other reasons, such as too costly of a process, product characteristics (like perishability), freight costs or lack of payback.

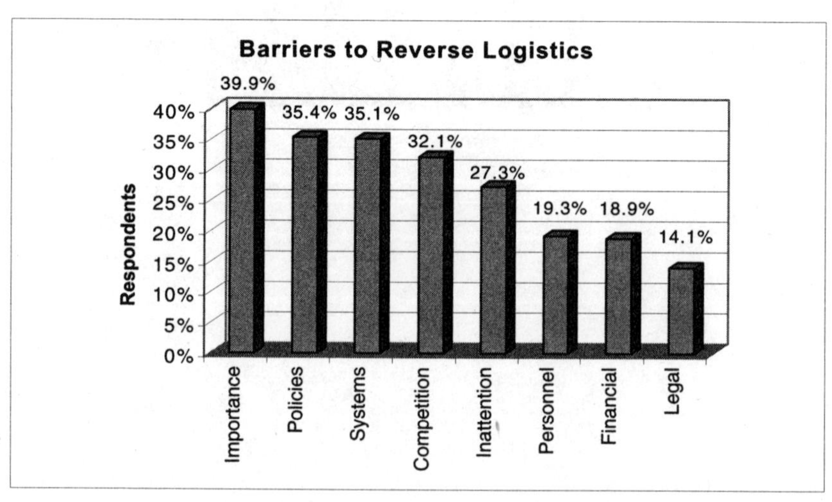

Question 12: What hardware and software technologies do you have installed, or plan to install, to assist your returns handling?

<u>Technologies</u>	<u>Respondents</u>
Bar codes	51.1%
Computerized return tracking	45.0%
Radio frequency (RF)	28.3%
Electronic data interchange (EDI)	27.0%
Computerized returns entry at most downstream point in supply chain	23.8%
Automated material handling equipment	19.3%

Other

5.14 percent of the participants mentioned other hardware and software technologies, such as automated Return Materials Handling processing, automated freight systems, virtual returns, and outsourcing.

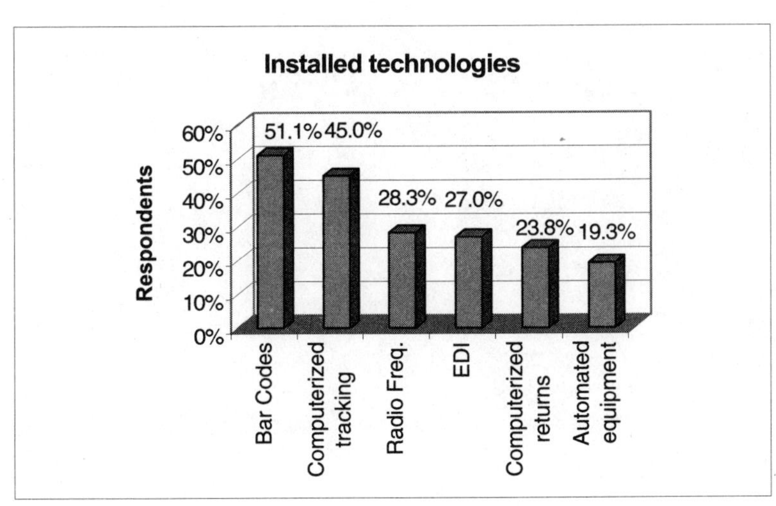

Question 13: How long is the returns processing cycle time for most of the products you handle?

Less than 1 day	4.8%
More than 1 day to 2 days	11.6%
More than 2 days to 1 week	25.6%
More than 1 week to 2 weeks	20.1%
More than 2 weeks to 1 month	23.2%
More than 1 month to 2 months	7.8%
More than 2 months	6.8%

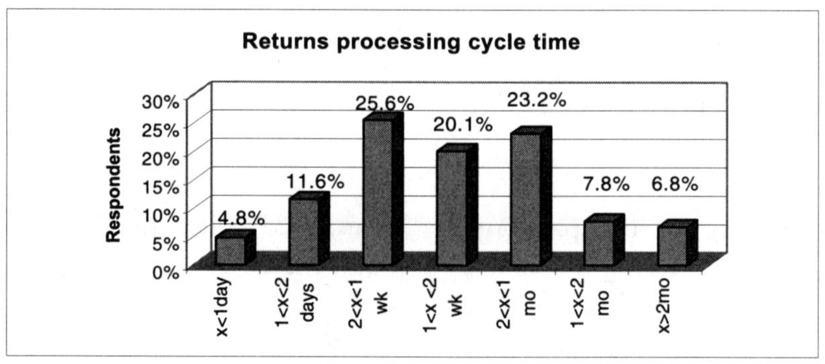

Question 14: What is your primary business?

Food	21.4%
Electronics and Computers	15.1%
General Merchandise	12.7%
Drugs, Health and Beauty Aids	11.5%
Automotive	9.9%
Apparel and Accessory	9.5%
Building, Materials, Hardware and Garden Supply	5.6%
Chemical	5.5%
Paper and Forest products	3.6%
Furniture, Home Furnishings, and Equipment	3.6%
Warehousing	1.6%

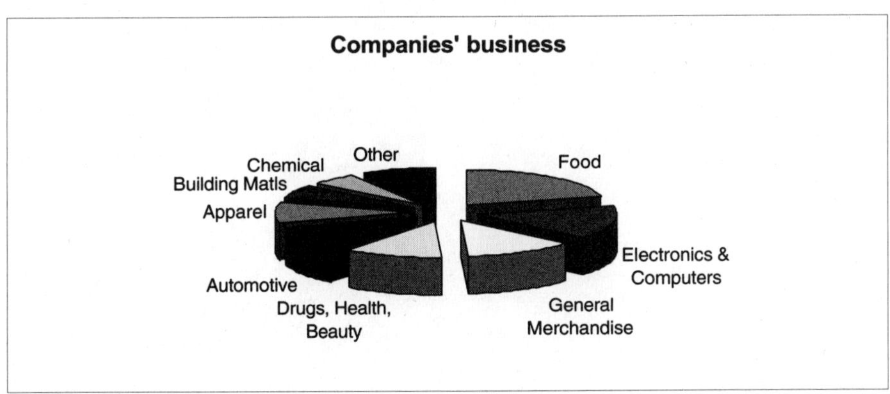

Other types of businesses mentioned are media and publishing, telecommunications, farm, and industrial and power equipment.

Question 15: On a scale of 1 to 7, with 1 being very unimportant, and with 7 being very important, rate the importance to your customers of each of the following in their decision to use you as a supplier:

	<u>Average</u>	<u>Standard deviation</u>
Cost Reduction	4.2%	4.4
Price	5.2%	1.7
Quality of Service	5.9%	1.5
Return Policies	3.7%	1.8
Speed of delivery	5.3%	1.8
Variety of products	5.1%	2.0

Question 16: What were the annual gross dollar sales of your business during the most recent fiscal year?

$5 Million or Less	0.7%
Over $5 to $10 Million	0.3%
Over $10 to $50 Million	3.4%
Over $50 to $100 Million	5.8%
Over $100 to $150 Million	4.8%
Over $150 to $200 Million	6.1%
Over $200 to $250 Million	3.1%
Over $250 to $500 Million	14.6%
Over $500 Million to $1 Billion	14.0%
Over $1 Billion	47.3%

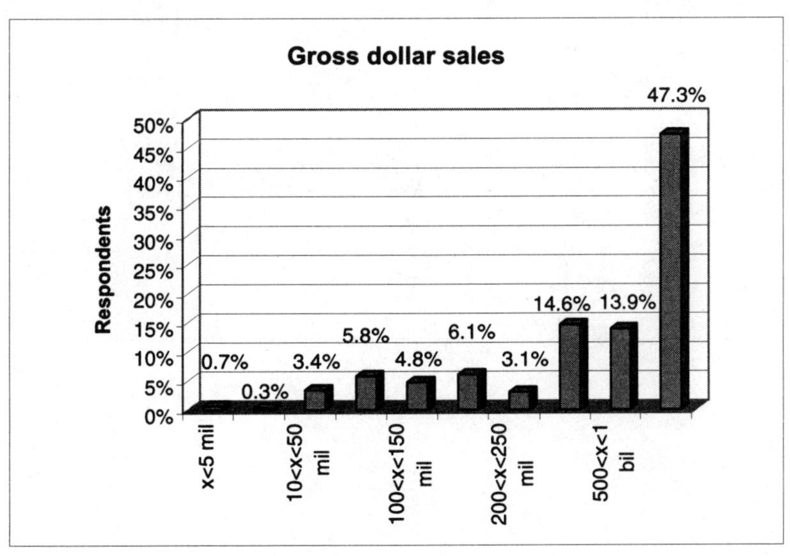

Question 17: How many people do you currently employ at this facility?

100 or fewer	16.8%
101 to 150	8.7%
151 to 250	9.1%
251 to 500	18.1%
501 to 1000	11.7%
Over 1,000	35.6%

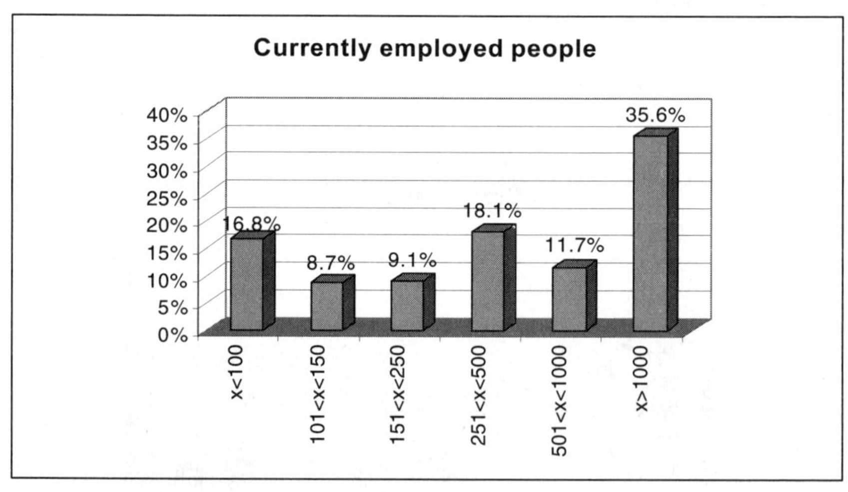

Question 18: In which of the following channel positions do you operate?

Manufacturer	64.0%
Wholesaler	29.9%
Retailer	28.9%
Service Provider	9.0%

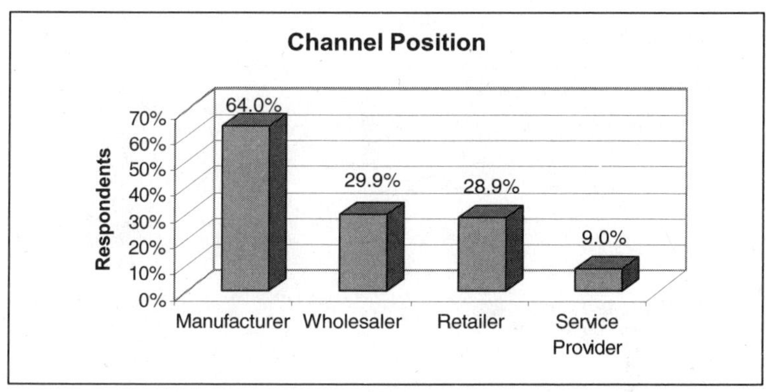

Appendix C: For More Information

Reverse Logistics Associations
Reverse Logistics Executive Council (RLEC)
Mail Stop 024
Reno, NV 89557
Phone: (775) 784-4912
Fax: (775) 784-1773
Web Site: www.rlec.org

Automotive Parts Rebuilders Association (APRA)
4401 Fair Lakes Court, Suite 210
Fairfax, VA 22033
Phone: (703) 968-2772
Fax: (703) 968-2878
Email: APRAMail@aol.com
Web Site: www.apra.org

International Reciprocal Trade Association
175 West Jackson Blvd., Suite 625
Chicago, IL 60604
Phone: (312) 461-0236
Fax: (312) 461-0474
Email: admin1@irta.net
Web Site: www.irta.net

Investment Recovery Association
5818 Reeds Road
Mission, KS 66202-2740
Phone: (913) 262-4597

Fax: (913) 161-0174
Email: ira@invrecovery.org
Web Page: www.invrecovery.org

Remanufacturing Industries Council International (RICI)
4401 Fair Lakes Court, Suite 210
Fairfax, VA 22033-3848
Phone: (703) 968-2995
Fax: (703) 968-2898
Email: sparker@rici.org
Web Page: www.rici.org

Logistics Associations
Council of Logistics Management
2805 Butterfield Road, Suite 200,
Oak Brook, IL 60523
Phone: (630) 574-0985
Fax: (630) 574-0537
Email: clmadmin@clm1.org
Web Site: www.clm1.org

International Warehouse and Logistics Association
1300 W. Higgins Road, Suite 111
Park Ridge, IL 60068-5764
Phone: (847) 292-1891
Fax: (847) 292-1896
Email: logistx@aol.com
Web Site: www.iwla.com

Environmental Legislation Information

Environmental Protection Agency
401 M Street, SW
Washington, DC 20460
Phone: (202) 260-2090
Email: public-access@epamail.epa.gov
Web Site: www.epa.gov

Duales System Deutschland AG
Frankfurter Straße 720-726
51145 Köln-Porz-Eil, GERMANY
Phone: 02203-937-0
Fax: 02203-937-190
Web Site: www.gruener-punkt-e/e/content/wie/wie00.htm

European Union policies and publications
Unit OP/3 'Publications'
2, rue Mercier
L-2985 Luxembourg
Fax: (352) 2929 44637
Email: idea@opoce.cec.be
Web Site: europa.eu.int/pol/en/env.htm

Packaging Recovery Organisation (PRO) Europe
Avenue de Tervuren 35, Etterbeek,
B – 1040 Bruxelles, BELGIUM
(An umbrella organization for national "Green Dot" programs in Europe.)

Take-Back & Environmental Information
Environmental Data Service (ENDS)
Finsbury Business Centre
40 Bowling Green Lane
London EC1R 0NE
ENGLAND
Email: post@ends.co.uk
Web Site: www.ends.co.uk
Phone: +44 (0) 171 814 5300
Fax: +44 (0) 171 415 0106

Electronic Product Recovery and Recycling Project (EPR2)
of the Environmental Health Center, a Division of
The National Safety Council
1025 Connecticut Ave. NW, Suite 1200
Washington, DC 20036
Phone: (202) 293-2270
Fax: (202) 293-0032
Web Site: www.nsc.org/ehc/epr2

Raymond Communication, Inc.
6429 Auburn Ave.
Riverdale, MD 20737-1614
(301) 345-4237
Fax: (301) 345-4768
Web Site: www.raymond.com
Email: michele@raymond.com

Reverse Logistics Providers

GENCO Distribution System
100 Papercraft Park
Pittsburgh, PA 15238
Phone: (800) 224-3141
Email: info@genco.com
Web Site: www.genco.com

Damagetrack
4045 University Parkway
Winston-Salem, NC 27106
Phone: (336) 896-7900
Fax: (336) 896-8190
Email: mgreenly@damagetrack.com
Web Site: www.damagetrack.com

EDI Transaction Sets

www.harbinger.com/resource/X12

Appendix D: EDI 180 Transaction Set

D.1 EDI Basics

Electronic Data Interchange (EDI) allows companies to exchange information electronically in a very compact, concise, and precise way. Because the transactions are compact and must follow strict standards, attempting to understand the language of these transactions is a study in minutia.

An EDI "exchange" is essentially the computer at one company making a phone call to the computer at another company. To allow everyone to speak the same language, these messages are sent in a standardized form, called "transaction sets." Messages about a request for quotation follow the 840 transaction set; messages about purchase orders follow the 850 transaction set. All in all, there are hundreds of transaction sets for different types of business situations.

Suppose one company (e.g., a retailer) wants to send a message to another company (e.g., a supplier) about a returned good authorization, the message must conform to the structure laid out in the "180 transaction set."

Within one interchange, several messages (called "transaction sets" in EDI terminology) about returns may need to be sent, in addition to other messages. All messages using the same transaction set are grouped together (into something called a "functional group.") For example, Figure D.1 shows how the contents of the phone call would be organized if the retailer wanted to ask about returning two products and submit four purchase orders.

Figure D.1
EDI Transaction structure

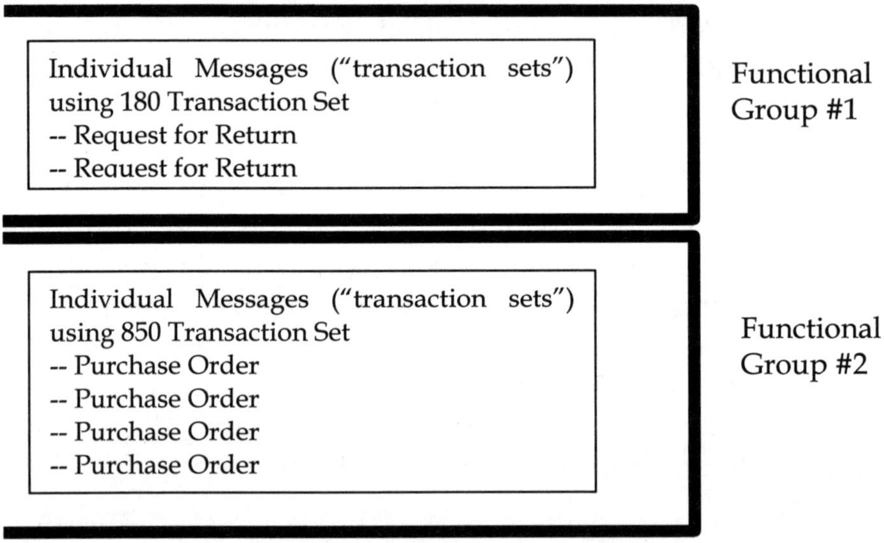

Each transaction set is made up of a number of pieces of information called "data segments." These are lines of the little codes that will be discussed in the next section. Fortunately, the end user generally does not have to be concerned about these codes. Using translation software, the purchase order information sitting in the retailer's information system is translated into an EDI transaction.

The data segments in a transaction set must conform to a standard order, which is shown in something called a transaction set "table." For example, the information about a return must follow the format in the 180 transaction set table, as shown in Appendix D.4. Not all of these data segments need to be present, but those elements that are present need to be in this sequence.

To give an example of how data elements are used, consider one segment (one "message") of the EDI transaction for a simple purchase order, as shown Table D.1.

At the start and end of the transaction set, there are data segments that signal that the segment is a purchase order. In between these markers is all of the relevant information for the purchase order.

Table D.1
Sample Purchase Transaction

```
ST*850*1                          Transaction set identifier
BEG*00*NE*00498765**010698        Beginning of the segment
PID*X*08*MC**Large Widget         Description of the product
PO1**5*DZ*4.55*TD                 Baseline Item Data
CTT*1                             Transaction Totals
SE*1*1                            End of the segment
```

The first line (the first "data segment") identifies that this segment will be a purchase order. The second segment indicates the beginning of the segment. The third segment gives information about exactly which product is to be ordered. Segment 4 contains information about the quantity and the price. Segment 5 tells how many line items have been ordered, and segment 6 signals the end of the segment.

The individual terms that make up each segment are called "data elements." Each segment starts with a data element that lets the supplier's computer know what the elements coming next represent. The "P01" on line 4 is how the supplier knows that the next few data elements will contain information about something called "Purchase Order Baseline Item Data." Asterisks are used to separate the various pieces of information that must go with each code. Because there are two asterisks in a row after the initial label, this means that there is a space for an optional data element, which has not been included because this information is not needed for the retailer's current transaction.

The second data element included, "5," indicates that the customer wants to order 5 units. The following "DZ" means that the units are in terms of dozens, so the customer wants 5 dozen products. The only way to know the meaning of "DZ" is to look it up in a table or with software. The fourth data element, "4.55" means that the price is $4.55. The last data element, "TD" means that the price given is the contract price per dozen. Again, the only way to know what each data element represents is to look it up in the layout of the transaction set.

Because some data elements are used by many transaction sets, each type of element has a number associated with it, (a "Data Dictionary Reference Number") for convenience. For example, the data element that tells what type of units the products are in is data element 355, the "Unit or Basis for Measurement Code." More information about the element can be found in a "Data Dictionary."

In some transaction sets, if optional data elements come at the end of the line, the line will terminate after the last value supplied.

D.2 EDI 180 Transaction Set

The EDI 180 transaction set "Return Merchandise Authorization and Notification" was designed to allow companies to exchange information about returns via EDI. It includes many data elements common to other transaction sets: Carrier Details, Address Information, etc. The 180 transaction set table is shown below in Section D.4.

The main data elements unique to the 180 transaction set involve the RDR (Return Disposition Reason) data segment. RDR is used to indicate the disposition of the item, the reason for the return, a description of the problem, and whether or not the item has been used, as shown in Table D.2. Several examples of the use of this data element are given at the end of this section.

Table D.2
RDR Segment Diagram Key

Sequence	Element	Name	Atr
01	1292	Returns Disposition Code	0
02	1293	Return Request Disposition Code	C
03	1294	Return Response Reason Code	C
04	352	Description	0
05	1073	Yes/No Condition or Response Code	0

The "Atr" column in Table D.2 indicates whether the element is mandatory, optional, or conditional, depending on whether the value is M, O, or C. A conditional element may be required, depending on what other elements have been used, and what values have been given for those element.

Data element 1292, "Returns Disposition Code" indicates how a contested item is to be disposed of. The possible values for 1292 are given in Table D.3.

Table D.3
1292 Returns Disposition Data Element

CODE	MEANING
CR	Consumer Return to Vendor
DI	Dispose
KA	Keep with an Allowance
KR	Keep and Repair
MW	Manufacturer Warranty Service

```
RA  Return with Authorization Number
RD  Request Denied
RF  Return for Factory Repair
RN  Return without Authorization Number
RP  Return Authorization Pending
RT  Ship to Third Party
SC  Ship to Third Party for Charitable
    Contribution
SD  Ship to Third Party for Disposal
```

Data element 1293, "Return Request Reason Code," is used to indicate why the returning party would like to return the item. Possible values for it are shown in Table D.4.

Table D.4:
1293 Return Request Reason Data Element

CODE	MEANING
CO	Customer Ordering Error
CV	Color Variance
DM	Defective Merchandise or Store Inspection
DP	Defective Packaging
DR	Defective Merchandise or Returned by Consumer
EI	Excess Inventory
EO	End of Season
EW	Excessive Wear
LP	Label Problem
NA	Not as Expected
OP	Outdated Packaging

```
PE   Price Error
PF   Poor Fit
PW   Poor Workmanship
SD   Short-Dated Product
SP   Shipped past Cancel Date
SR   Stock Reduction Agreement
ST   Style Problem
WG   Wrong Goods or Not Ordered
```

Data element 1294 is the "Return Response Reason Code." This is where the manufacturer is able to respond to the retailer, telling why the return was not authorized if it was not authorized. If the return was authorized, this code will be absent.

Table D.5
1294 Return Response Reason Data Element

CODE	MEANING
EW	Excessive Wear and Tear
FR	Freight or Retailer Damage
IN	Item not Defective
IO	Item as Ordered
MI	More Information Requested
NR	No Record of Original Sale
OS	Out of Season or Discontinued Line Item
PC	Pricing or Cost Difference
PR	Picture Requested
QD	Quantity Difference
RR	Repair or Refurbish
RT	Return Time Limit Exceeded or Beyond

<pre>
            Warranty Period
   SR   Sample Requested
   UI   Unidentifiable Item
</pre>

The complete180 transaction set is available on-line at
http://www.harbinger.com/resource/X12/.

Examples

RDR**DR**Y   Retailer asking to return product returned by a customer.

RDR*CR   Message sent from manufacturer instructing retailer to return the product to the vendor.

RDR**DM*EW   Retailer asked to return product because of defect, but manufacturer refused the return, saying that the problem is due to excessive wear and tear.

D.3 Criticisms of the 180 Set

Despite the fact that virtually every retailer or manufacturer in the U.S. could potentially use this transaction set, the set is barely used. Of the companies interviewed in the research project, which include most of the major retail chains in the U.S., none indicated that they were using the 180 set.

One common reason for not using the set is that the disposition and return reason codes are not sufficiently broad to cover all possible situations.

Additional Disposition Codes

From the research respondents, the following disposition codes would be helpful to have, in addition to the existing codes:

> Destroy on site
> Secure destruction (accompanied by security
>   personnel)
> Videotaped secure destruction
> Donate to charity locally
> Sell to broker (secondary markets)
> Exchange
> Resale
> Miscellaneous

For facilities that perform remanufacturing operations, the following terms would also be helpful:

> Rework
> Remanufacture
> Modify (Configuration or upgradable products)
> Repair

Return Reason Codes

For products being sent back, additional information about why the product is being sent back can increase the level of communication between retailer and supplier.

Repair / Service Codes

    Factory Repair – Return to vendor for repair
    Service / Maintenance

Order / Processing Codes

    Agent Order Error – Sales agent ordering error
    Internal Order Error – Incorrect internal ordering
    Entry Error – System processing order
    Shipping Error – Shipped wrong material
    Incomplete Shipment – Ordered items missing
    Wrong Quantity
    Duplicate Shipment
    Duplicate Customer Order
    Missing Part

Damaged / Defective

    Damaged – Cosmetic
    Dead on Arrival – Did not work
    Defective – Not working properly
    Defective – Distribution Center Inspection

Contractual Agreements

    Stock Adjustment – Rotation of Stock
    Obsolete – Outdated
    Seasonal

Other
>   Freight Claim – Damaged during shipment
>   Miscellaneous

D.4 Complete 180 Transaction Set Table

Any returns information sent via EDI must conform to the structure laid out in the transaction set table for the 180 transaction set, which is shown in Table D.6, below.

In each row, the second column "Seg" gives the data segment identifier for the data segment, followed by the name of the segment. The fourth column shows whether the segment is mandatory or optional. The fifth column (Max) shows the maximum number of these data segments that may be present. The final column (Loop) shows which segments may be repeated.

For example, lines 120 through 160 all give information about an address, and all have a "1" in the Loop column. This means that if the sender wants to include more than one address, each of these lines may be included. The "Loop Repeat-200" above line 120 indicates that as many as 200 address lines may be included in one transaction set in this way.

Table D.6
180 Transaction Set Table

180 - Return Merchandise Authorization and Notification Segments

Pos	Seg	Name	Req	Max	Loop
		TABLE 1			
010	ST	Transaction Set Header	M	1	
020	BGN	Beginning Segment	M	1	
030	RDR	Return Disposition Reason	O	1	
040	PRF	Purchase Order Reference	O	1	
050	DTM	Date/Time Reference	O	10	
060	N9	Reference Number	O	10	
070	PER	Administrative Communications Contact	O	2	
080	ITA	Allowance, Charge or Service	O	10	
090	PKG	Marking, Packaging, Loading	O	5	
100	TD1	Carrier Details (Quantity and Weight)	O	10	
110	TD5	Carrier Details (Routing Sequence/Transit Time)	O	10	
		LOOP ID-N1 Loop Repeat-200			1
120	N1	Name	O	1	1
130	N2	Additional Name Information	O	2	1
140	N3	Address Information	O	2	1
150	N4	Geographic Location	O	1	1
160	PER	Administrative Communications Contact	O	5	1
		LOOP ID-LM Loop Repeat-10			1
170	LM	Code Source Information	O	1	1
180	LQ	Industry Code	M	100	1
		TABLE 2			
		LOOP ID-BLI Loop Repeat-500			1
010	BLI	Baseline Item Data	O	1	1
011	N9	Reference Number	O	20	1
020	PID	Product/Item Description	O	5	1
030	RDR	Return Disposition Reason	O	1	1
040	ITA	Allowance, Charge or Service	O	10	1
050	PRF	Purchase Order Reference	O	1	1
051	AT	Financial Accounting	O	1	1
052	DTM	Date/Time Reference	O	15	1
053	DD	Demand Detail	O	100	1
054	GF	Furnished Goods and Services	O	1	1
055	TD5	Carrier Details (Routing Sequence/Transit Time)	O	5	1
		LOOP ID-LM Loop Repeat-10			21
056	LM	Code Source Information	O	1	21
057	LQ	Industry Code	M	100	21
		LOOP ID-N1 Loop Repeat-200			21
060	N1	Name	O	1	21
070	N2	Additional Name Information	O	2	21

080	N3	Address Information		O	2	21
090	N4	Geographic Location		O	1	21
100	PER	Administrative Communications Contact		O	5	21
		LOOP ID-QTY	Loop Repeat-1			21
110	QTY	Quantity		O	1	21
120	AMT	Monetary Amount		O	5	21
130	DTM	Date/Time Reference		O	10	21
140	N1	Name		O	1	21
		LOOP ID-LM	Loop Repeat-10			321
150	LM	Code Source Information		O	1	321
160	LQ	Industry Code		M	100	321
		LOOP ID-LX	Loop Repeat-1			321
170	LX	Assigned Number		O	1	321
180	N9	Reference Number		O	1	321
190	DTM	Date/Time Reference		O	10	321
200	N1	Name		O	1	321
		LOOP ID-LM	Loop Repeat-10			4321
210	LM	Code Source Information		O	1	4321
220	LQ	Industry Code		M	100	4321
230	SE	Transaction Set Trailer		M	1	

Glossary

"A" channel – the primary sales channel, carrying first quality goods that have not been available elsewhere.

Advance Ship Notice (ASN) – EDI transaction that informs users what, where, how, and when product is arriving.

Asset recovery – the classification and disposition of surplus, obsolete, scrap, waste and excess material products, and other assets, in a way that maximizes returns to the owner, while minimizing costs and liabilities associated with the dispositions.

"B" channel – secondary sales channel, for goods that have been through a reverse flow. Can carry first quality goods.

Barter companies – allow firms to get rid of unwanted inventories of first-quality and other goods, by trading for other products or for commodities such as airline tickets or advertising time.

Brokers – In reverse logistics, brokers are firms specializing in products that are at the end of their sales life. Often, willing to purchase any product, in any condition, given a low enough price. Often the customer of last resort for many returns.

Brown goods – electronics goods (such as computers, televisions, fax machines, and audio equipment).

Buy-out – when one manufacturer buys out a retailer's inventory of another manufacturer's product. This allows the buying manufacturer to replace its competitor's product with its own.

Cannibalization of demand – In reverse logistics, cannibalization of demand is when secondary market sales reduce sales in the "A" channel.

Cannibalization of parts – when parts or components are taken off of one item and used to repair or rebuild another unit of the same product.

Centralized Return Center (CRC) – a facility where a company's returns are processed.

Chargeback – a deduction from a vendor invoice for product return amount; sometimes occur without vendor permission.

Close-out liquidators – firms specializing in buying all of a retailer's product in some particular area; it usually happens when a retailer decides to get out of a particular area of business.

Controlled tip – a sanitary landfill where refuse is sealed in cells formed from earth or clay.

Core – a valuable and reusable part or subassembly that can be remanufactured and sold as a replacement part; often found in the automotive industry.

Core charge – the amount charged by a supplier on a remanufacturable product to encourage the consumer to return the defective item being replaced.

Design For Disassembly (DFD) – designing a product so it can be more easily disassembled at end-of-life.

Design For Logistics (DFL) – designing a product to function better logistically. Taking into consideration how the product will be handled, shipped, stored, etc.

Design For Manufacturing (DFM) – taking manufacturing concerns into account when designing a product, to enable easy manufacturing, cost effectiveness, or a higher standard of quality.

Design For Reverse Logistics (DFRL) – designing products so that their return flow functions better; designing reverse logistics requirements into product and packaging.

Disposition – how a product is disposed of, e.g. sold at an outlet, sold to a broker, sent to a landfill, etc.

Disposition cycle time – the duration of time from an item's initial return, to the item reaching its final disposition.

Duales System Deutschland (DSD) – the German organization responsible for collecting and recycling consumer packaging.

Electronic Data Interchange (EDI) – a system for business-to-business electronic communication.

Extended Producer Responsibility (EPR) – a requirement that the original producer of an item is responsible for ensuring its proper disposal.

Factory-renewed – a product that has been refurbished by the manufacturer; typically carries a full new-product warranty.

Footprint – building size, in square feet. A large footprint store requires a large number of square feet.

Gatekeeping – the screening of products entering the reverse logistics pipeline.

Gray market – products sold through unauthorized dealers or channels; generally do not carry a factory warranty.

Green Dot – a symbol on packaging sold in Germany that indicates that the product is eligible to be recycled through the Duales System Deutschland.

Green logistics – attempts to measure and minimize the ecological impact of logistics activities.

High learning products – items that require education or instruction before being able to operate; a computer, for example.

Insurance liquidators – secondary market companies specializing in buying products damaged in shipment and declared as losses by insurance companies.

Investment recovery – see asset recovery.

Irregular – products that do not meet the standards for first-quality product, perhaps for cosmetic reasons, but which generally still satisfy most of the basic performance requirements.

Job-out liquidators – secondary market companies specializing in buying end-of-season products from retailers.

Landfill – a controlled environment for burying municipal solid waste.

Leachate – water that seeps through a landfill, picking up pollutants as it travels.

Liquidator – a secondary market company that buys product that has reached the end of its sales life in the "A" channel.

Lift – see buy-out.

Logistics – the process of planning, implementing, and controlling the efficient, cost-effective flow of raw materials, in-process inventory, finished goods, and related information from the point of origin to the point of consumption for the purpose of conforming to customer requirements.

Made-for-outlet – products made especially to be sold at outlet stores; generally of slightly lower quality that "A" channel products.

Marketing returns – unsold product a supplier has agreed to take back from the retail customer; usually overstocks; can be the result of product shipped to the retailer with the understanding that sales are guaranteed.

Municipal Solid Waste (MSW) – garbage generated by residences and small businesses.

Non-defective defectives – when customers return a product claiming it to be defective, when in fact, the problem is not with the product, but often with the customer's ability to properly operate the product.

Non-defective returns – a non-defective defective returned by a customer.

Outlet sales – products sold at an "outlet" store; typically irregular or off-season products.

Overstock – excess inventory; may be from ordering too much, order cancellations, or product's failure to sell.

Point Of Sale (POS) –the point where ownership of the product transfers to the customer.

Point-Of-Sale (POS) registration – collecting customer registration information for warranty purposes at the time the product is sold.

Partial returns credit – giving a customer a partial refund for a product because not all components of the product are present.

Prebate – providing a discounted purchase price on a product linked to the promise not sell the product to a remanufacturer at the end of its life; paying the customer at the time of purchase for returning the product at end-of-life.

Preselling – contracting ahead of time (during the selling season) with a job-out company to purchase all remaining product at the end of the season.

Primary packaging – the first level of product packaging; for example, the tube that toothpaste is packed in, or a bottle that contains beer.

Producer pays – the principle that the manufacturer should pay for ensuring the recycling and proper disposal of product at end-of-life.

Reclaim materials –see recycling.

Reclamation centers – centralized processing facilities for returns; term used widely in the grocery industry.

Reconditioning – when a product is cleaned and repaired to return it to a "like new" state.

Recycling – when a product is reduced to its basic elements, which are reused.

Refurbishing – similar to reconditioning, except with perhaps more work involved in repairing the product.

Remanufacturing – similar to refurbishing, but requiring more extensive work; often requires completely disassembling the product.

Re-returns –when a customer tries to return for full price a product that was sold as a returned product.

Resell – when a returned product may be sold again as new.

Restocking fee – a charge to the consumer for accepting their returned product.

Return abuse – when a customer tries to return a product at a chain other than where they bought it, or for a price higher than what they paid for it, or after the warranty period has expired.

Returnable tote – transport packaging that can be used multiple times to move materials between or within facilities.

Return Authorization (RA) – authorization to return a product to a supplier.

Return Material Allowance (RMA) – authorization to return a product to a supplier.

Returns – products for which a customer wants a refund because the products either fail to meet his needs or fail to perform.

Returns allowance – the quantity of product that a customer is allowed to return; usually calculated as a percentage of total purchases.

Returns center – same as centralized return center.

Return to supplier – returning damaged products or customer returns to the vendor from whom they were purchased.

Return To Vendor (RTV) – same as return to supplier.

Reusable tote – same as returnable tote.

Reuse – using a product again for a purpose similar to the one for which it was designed.

Reverse distribution – the process of bringing products or packaging from the retail level through the distributor back to the supplier or manufacturer.

Reverse logistics – the process of planning, implementing, and controlling the efficient, cost-effective flow of raw materials, in-process inventory, finished goods, and related information from the point of consumption to the point of origin for the purpose of recapturing value or proper disposal.

Radio Frequency Identification (RFID) – a technology in which a tag is attached to each item which broadcasts a unique, low-frequency radio signal.

Return Material Authorization (RMA) – permission to return a product to a supplier.

Rotable parts – using a closed loop of repairable products; when a customer sends in a broken product, a repaired product is sent, and the customer's product is repaired and stored to be sent to another customer.

Salvage – when a product is sold to a broker or some other low-revenue customer.

Sanitary landfill – a landfill scientifically designed to prevent groundwater contamination from leachate.

Secondary market – a collection of companies that specialize in selling products that have reached the end of their selling season in the "A" channel.

Secondary packaging – the second level of product packaging; for example, the box that contains a tube of toothpaste, or the carton that holds six bottles of beer.

Secure disposal – requiring a company to destroy the product under the supervision of a security guard to ensure the product is destroyed.

Secure returns – a reverse logistics process designed to minimize leakage of product; secure returns processes are designed to eliminate shrinkage and unwanted product disposition.

Source reduction – reducing usage of resources at the point of generation or production.

Supply chain position – the position in the channel that the firm occupies; this position could be manufacturer, wholesaler, distributor, retailer, or combinations of these.

Take-back – requiring manufacturers to collect product at end-of-life to reclaim materials and dispose of properly.

Tipping fees – the cost of disposing of one ton of garbage in a landfill.

Transport packaging – packaging used for transporting products from manufacturers to distributors or retailers.

Two-dimensional bar coding – a bar coding technology that allows much more information to be stored in a given space; instead of a single row of line, the bar code label consists of a two-dimensional grid of dots.

White goods – household appliances such as washers, dryers, and refrigerators.

Zero returns – manufacturer never takes possession of returns. Destroyed in the field by retailer or third party.

Endnotes

Chapter 1

1   Bob Delaney, *Ninth Annual State of Logistics Report*, (St. Louis, MO: Cass Logistics, 1998).
2   Automotive Parts Rebuilders Association, *Rebuilding/Remanufacturing: Saving the World s Environment*, (Fairfax, VA: Automotive Parts Rebuilders Association, 1998).
3   Lee, Louise, "Without a Receipt, You May Get Stuck With That Ugly Scarf," *Wall Street Journal*, November 18, 1996, sec. 1, p. 1.
4   Annette Spence, "Hannadowns," *Sky Magazine*, May 1998, pp. 107-111.
5   Richard L. Dawe, "Reengineer Your Returns," *Transportation and Distribution*, August 1995, pp. 78-80.

Chapter 2

1   Martijn Thierry et al., "Strategic Issues in Product Recovery Management," *California Management Review* 37, no. 2, Winter 1995, pp. 114-135.
2   Ibid.

Chapter 3

1   "Outlet Malls: Do They Deliver the Goods?," *Consumer Reports*, August, 1998, pp. 20-25.
2   Ibid.
3   Interview with Herb Shear, GENCO Logistics System, Pittsburgh, PA, March 20, 1997.

Chapter 4

1   Edward A McBean, Frank A. Rovers, and Grahame J. Farquhar, *Solid Waste Landfill Engineering and Design*, (Englewood Cliffs, New Jersey: Prentice Hall, 1995).
2   Ibid.

3   Ibid.

4   Walter H. Corson, *The Global Ecology Handbook,* (Boston: Beacon Press, 1990).

5   O.P Kharbanda and E. A. Stallworthy, *Waste Management: Towards a Sustainable Society,* (New York: Auburn House, 1990).

6   Environmental Protection Agency, *Characterization of Municipal Solid Waste in the United States: 1994 Update Executive Summary,* (Washington, D.C.: U.S. Environmental Protection Agency, Nov. 1994).

7   Environmental Protection Agency, *List of Municipal Solid Waste Landfills,* (Washington, D.C.: U.S. Environmental Protection Agency, 1995).

8   Ibid.

9   Environmental Protection Agency, *Municipal Solid Waste Fact Book Internet Version,* (Washington, D.C.: U.S. Environmental Protection Agency, 1997).

10   Environmental Protection Agency, *Characterization of Municipal Solid Waste in the United States: 1996 Update,* (Washington, D.C.: U.S. Environmental Protection Agency, 1996).

11   Pennsylvania Department of Environmental Protection, *Average Landfill Tipping Fees for Municipal Solid Waste in Pennsylvania,* (Harrisburg, PA: Pennsylvania Department of Environmental Protection, 1996).

12   Jeff Bailey, "Arizona has Plenty of What Oceanside Needs and Vice Versa: So a Swap is Arranged—Sand to Spread on the Beach For Loads of Garbage," *Wall Street Journal,* March 4, 1997, sec. 1, p. A1.

13   Ibid.; and Mark Lifsher, "Ward Valley Waste Dump Isn't Economical or Needed, Study Says," *Wall Street Journal,* December 3, 1997.

14   Stacy Kravetz, "These People Search for Cup That Suits the Coffee it Holds," *Wall Street Journal,* March 24, 1998, sec. 1, p. A1.

15   David Wallechinsky and Irving Wallace, *The People's Almanac* (New York : Doubleday & Company, Inc, 1975), p. 917.

16   William Rathje and Cullen Murphy. *Rubbish: The Archaelology of Garbage,* (New York: Harper Collins, 1992).

17   EPA Fact Book, 1997.

18 Institute of Scrap Recycling Industries, *Recycling: The Economically and Environmentally Intelligent Alternative to Landfilling and Incineration,* (Washington, D.C.: Institute of Scrap Recycling Industries, 1998).

19 EPA Fact Book, 1997.

20 Edward A. McBean, et al.

21 EPA Fact Book, 1997.

22 California Integrated Waste Management Board, *Solid Waste Landfilling Data,* (Sacramento, CA: California Integrated Waste Management Board 1997).

23 EPA Fact Book, 1997.

24 Ibid.

25 Robert Steuteville, Nora Goldstein, and Kurt Grotz, "The State of Garbage in America," *BioCycle,* 34, June 1993, pp. 32- 37.

26 Ibid.

27 Brian Cunningham and James R. Distler, "Reverse Logistics Shock," in Council of Logistics Management, *Annual Conference Proceedings,* (Chicago, IL: Council of Logistics Management, 1997), pp. 423-426.

28 Advanced Recovery, *CRT Recycling,* (Belleville, NJ: Advanced Recovery, 1998).

29 Cheryl L. McAdams, "Resurrecting the Computer Graveyard," *Waste Age,* Feb. 1995.

30 Steve Lohr, "Recycling Answer Sought for Computer Junk," *New York Times,* April 14, 1994.

31 Steve Anzovin, "The Green PC Revisited," *Dallas/Fort Worth Computer Currents,* December, 1997, pp. 38-45.

32 Daniel Machalaba, "Hitting the Skids: As Old Pallets Pile Up, Critics Hammer them as New Eco-Menace," *Wall Street Journal,* April 1, 1998, sec. 1, p. A1.

33 Steueteville, et al., p. 34.

34 National Wooden Pallet & Container Association, *Recycling Solutions for Pallet Disposal,* (Arlington, VA: National Wooden Pallet & Container Association, 1996).

35 Raymond Communications, *Transportation Packaging and the Environment,* (College Park, MD: Raymond Communications, 1997).

[36] Minnesota Mining and Manufacturing Corp., *3M Gives Business an Eco-Advantage,* (St. Paul, MN: Minnesota Mining and Manufacturing Corp., 1997).

[37] Tom Andel, "It's a Two-Way Stream: Logistics Packaging is Flowing Back to Suppliers for Re-use, and Recycling," *Transportation and Distribution,* December, 1997, pp. 81-91.

[38] Raymond Communications, 1997.

[39] Lori Kampschroer, James Rust, James G. Thompson, Diana Twede, and Steve Vagsnes, "Do Returnable Containers for Large Finished Goods Make Sense?," Council of Logistics Management Annual Conference, October 21, 1996.

[40] Diana Twede, "Do Returnable Containers for Large Finished Goods Make Sense? Returnable Packaging Considerations" in Council of Logistics Management, *Annual Conference Proceedings,* (Orlando, FL: Council of Logistics Management, 1996), pp. 585-587.

[41] "Consumer Polystyrene Recycling Faces Challenges," *State Recycling Laws Update,* Oct.-Nov. 1997, p. 6.

[42] Ibid.

[43] Ibid.

[44] Ibid.

[45] Ebora Vrana, "EarthShell IPO: Will Investors Contain their Enthusiasm?," *Los Angeles Times,* March 13, 1998, sec. 2, p. B1.

[46] Ibid.

[47] Amy Zuckerman, "Recyclable Containers: Recycling Packaging Materials Became a Trend 'Because it Costs so Much More to Take Out the Garbage'," *Traffic World,* February 17, 1997.

[48] A.R. van Goor and D.C. Loa, *Research on the RVT Supply Chain Model for Dry Grocery Goods,* (Amsterdam: Amsterdam Free University, Faculty of Economic Science and Business Administration, Department of Logistics, 1995).

[49] Gary A. Davis, Catherine A. Wilt, Patricia S. Dillon, and Bette K. Fishbein, "Extended Product Responsibility: A New Principle for Product-Oriented Pollution Prevention." (Washington, D.C.: U.S. Environmental Protection Agency, Office of Solid Waste, June, 1997).

[50] Ibid.

[51] Steueteville, et al., p. 35.

52 Colleen Mizuki, et al., *Electronics Industry Environmental Roadmap: Electronics Products Recycling Study*, (Austin, TX: Microelectronics and Computer Technology Corporation, 1996).

53 Microelectronics and Computer Technology Corporation, *The University of Texas and MCC Collaborate to Develop Comprehensive Plan for Electronics Products Recycling*, (Austin, TX: Microelectronics and Computer Technology Corporation, 1997).

54 "Results from Residential End-of-Life Electronics Collection Pilots are Forthcoming," *EPR2 Update*, [Washington, D.C.: The Electronic Product Recovery and Recycling Project of the Environmental Health Center, a division of The National Safety Council], Winter, 1998.

55 *Wall Street Journal*, "Business Beat: Big Blue Goes Green," July 10, 1997, sec. 1, p. A1.

56 G. Pascal Zachary, "Why We Can't Part with those Vintage PCs," *Wall Street Journal*, July 2, 1997, sec. 2, p. B1.

Chapter 5

1 Duales System Deutschland, *Why do companies become licensees of the Dual System?*, (Cologne, Germany: Duales System Deutschland AG, 1997).

2 Frank Ackerman, *Why do We Recycle?: Markets, Values, and Public Policy*, (Washington, D.C.: Island Press, 1997).

3 Ibid., 1998.

4 Duales System Deutschland, *Packaging Recycling: Techniques and Trends*, (Cologne, Germany: Duales System Deutschland AG, 1998).

5 Ibid.

6 "German Packaging Recovery Rate up in 1997, *Environment Daily*, [London], May 25, 1997.

7 Robert Steuteville, Nora Goldstein, and Kurt Grotz, "The State of Garbage in America," *BioCycle*, 34, June 1993, p. 33.

8 Ackerman.

9 Steueteville, et al., p. 35.

10 Ackerman.

11 Ibid.

12 Matthew Gandy, *Recycling and the Politics of Urban Waste*, (New York: St.Martin's Press, 1994).

13   Ackerman.

14   Gandy.

15   Raymond Communications, *Getting Green Dotted: The German Recycling Law Explained in Plain English*, (Riverdale, MD: Raymond Communications, 1998).

16   "German Packaging Recovery System Challenged," *Environment Daily*, [London] May 5, 1998.

17   Raymond Communications, *Transportation Packaging and the Environment*, (College Park, MD: Raymond Communications, 1997).

18   Andreas von Schoenberg and Sascha Kranendonk, *DSD: Industry-run packaging waste reduction system* (Wuppertal, Germany: Wuppertal Institute, 1995).

19   "TV's Too Expensive to Collect, Study Finds," *Recycling Laws International*, Sept. 1997, pp. 5-6.

20   "Progress Made on EU Electronics Waste Plan," *Environment Daily*, [London] April 28, 1998.

21   "EC Moves on Electronics Waste Directive," *Recycling Laws International*, Sept. 1997, p. 5.

22   *Environment Daily*, "Progress Made on EU Electronics Waste Plan."

23   "Electronics Recycling Discussed by EU States," *Environment Daily*, [London] May 27, 1998.

24   "Firms Criticize Draft EU Electrical Waste Law," *Environment Daily*, [London] June 23, 1998.

25   "EU Retailers Oppose Electronics Take-Back Plan," *Environment Daily*, [London] June 29, 1998.

26   "Norway to Require Electronics Waste Take-Back," *Environment Daily*, [London] March 16, 1998.

27   "Industries Query Viability of Take-Back Law," *Environment Daily*, [London] May 7, 1998.

28   *Recycling Laws International*, "TV's Too Expensive to Collect, Study Finds."

29   "Netherlands forces Appliance Take-Back," *Environment Daily*, [London] March 19, 1998; and *Environment Daily*, "Industries Query Viability of Take-Back Law."

30   "EU end-of-life Vehicles Proposal Released," *Environment Daily*, [London] July 7, 1997.

31 "EU Car Firms Oppose End-Of-Life Vehicle Plan," *Environment Daily*, [London] July 21, 1997.

32 Ibid.

33 "Car Dismantlers Support EU Recycling Plan," *Environment Daily*, [London] April 21, 1997.

34 "UK Voluntary Car Recycling Deal Struck," *Environment Daily*, [London] July 15, 1997.

35 Auto Recycling Nederland, *Environmental Report: Facts and Figures*, (Amsterdam: Auto Recycling Nederland BV, 1996).

36 "Sweden Requires Free Take-Back for New Cars," *Environment Daily*, [London] October 27, 1997.

37 "Battery Directive Would Ban Ni-cd's," *Recycling Laws International*, September 1997, p. 5.

38 Varta AG, *Varta Sets Out its Position in Response to the Upcoming German Battery Order*, (Hannover, Germany: Varta AG, April 30, 1997).

39 "Battery Take-Back Moving," *Recycling Laws International*, September 1997, p. 5.

40 "Sweden: Industry Wrist-Slap," *Recycling Laws International*, September 1997, p. 5.

Chapter 6

1 Richard J. Kish, "The Returns Task Force," *1997 Book Industry Trends*, in Chapter 1: "Issue of the Year: Returns" (New York: Book Industry Study Group, 1997).

2 G. Bruce Knecht, "Macmillan's Order to Use Wholesalers Angers Booksellers," *Wall Street Journal*, September 5, 1997, sec. 2, p. B4.

3 I. Jeanne Dugan, "Boldly Going Where Others are Bailing Out," *Business Week*, April 6, 1998, p. 46.

4 Elizabeth Lesly Stevens and Ronald Grover, "The Entertainment Glut," *Business Week*, February 16, 1998, p. 93.

5 Kish.

6 Ibid.; and Doreen Carvajal, "Returns are Swamping the Publishing Industry," *New York Times*, August 1, 1996.

[7] Hardy Green, "Superstores, Megabooks—and Humongous Headaches: With sales flat and returns piling up, many publishers are smarting," *Business Week,* April 14, 1997, pp. 92-94.

[8] Kish.

[9] Doreen Carvajal, "Publishers Seek Advice from Bookstore Chains," *New York Times,* August 12, 1997; and Barbara Carton, "Bookstore Survival Stunts Have Scant Literary Merit," *Wall Street Journal* June 3, 1997, sec. 2, p. B1.

[10] Cindy Hall and Dave Merrill, "USA Snapshots: The Word on Discount Books," *USA Today,* October 9, 1997.

[11] Bruce G. Knecht, "Chain Reaction: Book Superstores Bring Hollywood-Like Risks to Publishing Business," *Wall Street Journal,* March 24, 1997, sec. 1, p. A1.

[12] Cynthia Crossen, "Put a Pen to Paper These Days, You'll be a Published Author," *Wall Street Journal,* January 10, 1996, sec. 2, p. B1.

[13] Dugan.

[14] "Carnegie Mellon University Updates Computer Disposition Data," *EPR2 Update,* [Washington, D.C.: The Electronic Product Recovery and Recycling Project of the Environmental Health Center, a division of The National Safety Council], Summer, 1997.

[15] Raju Narisetti, "Printer Wars: Toner Discount Incites Rivals," *Wall Street Journal,* April 10, 1998, sec. 2, p. B1.

[16] Institute of Scrap Recycling Industries, *Recycling: The Economically and Environmentally Intelligent Alternative to Landfilling and Incineration,* (Washington, D.C.: Institute of Scrap Recycling Industries, 1998).

[17] Willam P. Steinkuller, "Recycling or Reuse: The Long and Short of it." *Ward s Auto World* 30, April 1994, p. 16.

[18] Drew Winter, "Ship Ahoy! Automakers Sail Into Recycling," *Ward's Auto World* 29, September 1993, p. 64.

[19] Gene Bylinski, "Manufacturing for Reuse," *Forbes* 131, February 6, 1995.

[20] Ibid.

[21] Winter, p. 64.

[22] Lindsay Brooke, "The Recyclability Gap," *Automotive Industries* 174, February 1994.

23 Warren Brown, "Chrysler Close to Turning Recyclables Into a Car," *Washington Post,* September 9, 1997, sec. 3, p. C3.

24 Tom Werner, "Blue Chip Products Isn't Blue–It's in the Chips," *Philadelphia Business Journal* 11, February 15, 1993.

25 "Ford Picks 'Core' Carrier," *Purchasing* 119, July 13, 1995, p. 101.

26 "Auto Parts Remanufacturer Tunes up with Symbol/True Data," *Industrial Engineering* 25, October 1993, p. 34.

Index

Dale Rogers, Ph.D.

Dale Rogers is the Director of the Center for Logistics Management and an Associate Professor of Marketing and Logistics at the University of Nevada. He received his Ph.D., MBA, and undergraduate degree at Michigan State University.

Prior to entering the doctoral program, Dale worked as a Project Director and Director of Marketing for a logistics consulting and software company. He has also been employed in material management and manufacturing at the Harris Corporation, and as an instructor with the Oldsmobile Division of General Motors.

Dale has published in several logistics journals and is a co-author of two previous books on logistics.

Dale is married to Deana Rogers. They have five children: Zachary, Benjamin, Tre, Samuel, and Jaesa.

Ron Tibben-Lembke, Ph.D.

Ron Tibben-Lembke is an Assistant Professor of Logistics at the University of Nevada, and Director of the Summer Logistics Internship Program. He received his Ph.D. and MS degrees in Industrial Engineering from Northwestern University, and his undergraduate degree in Mathematics and Computer Science from St. Olaf College in Northfield, Minnesota.

Ron has research interests in many areas of logistics. In addition to reverse logistics, these include: total cost of ownership, design of distribution networks, facility location and design, supply contracts, and postponement. Ron has worked with General Motors and ALRO Steel on designing distribution systems and with other companies in redesigning their facilities.

Ron is married to Jennie Tibben-Lembke. Their first child, Grace Johanna, was born on September 13, 1998.